FINITE ELEMENT METHOD USING BASIC PROGRAMS

D.K.Brown

B.Sc., Ph.D., C.Eng., F.I.Mech.E.

Senior Lecturer in Mechanical Engineering,
University of Glasgow
and
Director, Fracture Mechanics Consultants Ltd.

Surrey University Press

Distributed in the USA by
Chapman and Hall, New York

Published by Surrey University Press.
A member of the Blackie Group,
Bishopbriggs, Glasgow G64 2NZ and
Furnival House, 14-18 High Holborn,
London WC1V 6BX.

Distributed in the USA by
Chapman and Hall
in association with Methuen, Inc.,
733 Third Ave., New York, N.Y. 10017.

British Library Cataloguing in Publication Data

Brown, David K.
An introduction to the finite element method using BASIC programs.
1. Finite element method – Data processing
2. Engineering mathematics – Data processing
3. Basic (Computer program language)
I. Title
620'.001'515353 TA347.F5

ISBN 0-903384-46-9
ISBN 0-903384-45-0 Pbk

Library of Congress Cataloging in Publication Data

Brown, D. K.
An introduction to the finite element method using BASIC programs.
Bibliography: p.
Includes index.
1. Finite element method – Data processing. 2. Microcomputers – Programming.
3. BASIC (computer program language). I. Title.
TA347.F5B76   1984   624.1'71'01515353   84-241
ISBN 0-412-00571-9 (Chapman and Hall)
ISBN 0-412-00581-6 (Chapman and Hall: pbk.)

Filmset by Presentia Art
Printed in Great Britain by Bell and Bain Ltd., Glasgow.

Preface

The past 30 years have seen the parallel developments of high-speed low-cost computers and the finite element method. The advent of the powerful desk-top microcomputer now means that practising engineers have available to them in their offices previously unthought-of analytical capabilities. The techniques which a decade or two ago were still in the realms of research and development are now available as tools for everyday use. The teaching of finite elements to undergraduates and to practising engineers in post-experience courses is commonplace. The origins of finite elements lie in the structural field, which is the main concern of this book, but the technique has been applied to many other fields.

This book is an introduction to finite elements in simple two-dimensional structures. The approach starts with a redundant pin-jointed structure and solves it in the traditional longhand method, which derives from basic mechanics. A subtle change of formulation leads to the development of the pin-jointed element and the use of the stiffness method of matrix structural analysis. It is hoped that by this analytical device both students and engineers will find the key to the finite element techniques. The generalization to rigid-jointed two-dimensional frames then follows and both chapters have self-contained programs appended – **PJFRAME** and **PLFRAME**.

A different approach is taken in introducing the two other finite elements – the constant strain triangle for plane stress/strain and the rectangular element for the bending of thin flat plates. Both require the introduction of guessed displacement functions and the use of the principle of virtual work to optimize the guess. Like the frames, two self-contained programs, **FEP** and **FEPB**, are appended to the respective chapters.

All four programs are written in BASIC and full listings are given along with the output from sample solutions run with them. It is hoped that the BASIC used is reasonably machine-independent in order that the programs may be run on most micros.

One of the main benefits of running finite elements on micros is that the user is in full control of the machine, which may be switched off and the program easily re-loaded if difficulties arise! The fear of being 'locked in' to a large mainframe computer often deters engineers from use of large finite element packages. However once the reader has become acquainted with the method and programming techniques, it is expected that the use of such packages as PAFEC, MARC, FLASH2, ABAQUS, NASTRAN, etc., will not be such a daunting prospect.

As part of an undergraduate curriculum, it is often a good idea to introduce the technique over two or three years. The chapter on pin-jointed frames needs little basic mechanics and mathematics but allows an early introduction to the stiffness method. The development of mechanics to beam problems, and mathematics to further matrix algebra, allows the introduction of the beam element in **PLFRAME.** Finally the theories of elasticity and plates complement the introduction of the remaining two elements in a final-year undergraduate course.

The standardization of the programs and in particular the careful formatting of output was done by Mr David A. Pirie of the Department of Aeronautic and Fluid Mechanics at the University of Glasgow. He has also summarized the BASIC statements in Appendix B and commented on machine-dependent aspects.

D.K.B.

Contents

Acknowledgements

This book has its origins in the engineering curriculum at the University of Glasgow. In the early 1970s, Dr A.C. Mackenzie and Mr T.H. Cain introduced beam elements into the structures course. A debt of gratitude is due to both of them for the meticulous care that was put into presenting this technique to undergraduates. The material for the chapter on the constant strain triangle originates from a post-experience course run about the same time by Dr Mackenzie and the late Dr James Orr.

The author also acknowledges the support of the University of Glasgow in the resources it has made available for the development of the programs over several years. Finally, I extend my thanks to Miss Marilyn Dunlop, who prepared the manuscript, and to Mr Richard Appleton who assisted in proof reading.

To Caroline

1 Introduction

In the past 30 years structural analysis has changed dramatically with the advent, and then the enormous expansion, of the power of computers. The development of programs for specific applications led to the development of large packages, which took even larger computers to store. Whereas these large packages do undoubtedly have their place, in general the majority of industrial uses require a smaller and more efficient package, or even individual programs. The development of computers has now made available microcomputers, which have the capacity to run individual structural analysis programs and in the near future will be able to store and run larger packages. The language of the micros is usually BASIC and so the four programs presented here are written in BASIC and are suitable for running on Commodore PET 4000 series computers. It should be noted that with minor alterations the programs could be run on many different micro systems, such as the APPLE. Appendix B details the modifications.

An elastic structure can be thought of as a spring with stiffness (or flexibility) or as an assemblage of springs, which connect all major points or nodes in a structure to all other nodes. Just as a spring has a stiffness or flexibility coefficient, the structure, as an assemblage of springs, will have a complete set or matrix of such coefficients, which is usually called the stiffness or flexibility matrix of the structure. When each individual part of a structure is analysed there are derived several relations between applied loads on this part and displacements and rotations within the part – in other words a set of stiffness or flexibility coefficients for the part itself. Thus for each part or element of a structure a matrix of coefficients exists and these could be calculated with respect to the element's own coordinate system or the coordinate system defining the structure, sometimes referred to as the global coordinate system. The process of transferring the local stiffness or flexibility coefficients to the structural matrix is achieved by using force equilibrium or slope and displacement compatibility at the nodes where two or more elements meet. Such equilibrium or compatibility sums not only include inter-element interactions but also must include externally imposed load or displacement conditions.

There results a set of linear algebraic simultaneous equations, when linear elastic conditions are used. Now, for a single spring of stiffness k, the single linear equation is

$$k.u = p \qquad (1.1)$$

or if f is the flexibility then

$$f.p = u \qquad (1.2)$$

If a known value of force p is applied in (1.1), the displacement u can be found by dividing through by k, or multiplying both sides by k^{-1} (the 'inverse' of k)–

$$u = k^{-1} p$$

Similarly in (1.2) if a known value of u is experienced by the spring, the reacting force can be calculated by multiplying u by the inverse of f:

$$p = f^{-1} u$$

Of course, the inverse of stiffness is flexibility, $k^{-1} = f$, and vice versa, $f^{-1} = k$. In a structure, the single stiffness or flexibility is replaced by a matrix of coefficients $[K]$ or $[F]$ and the displacement and force by a vector of nodal displacements $\{U\}$ or nodal forces $\{P\}$ Thus

$$[K]\{U\} = \{P\} \qquad (1.3)$$

or

$$[F]\{P\} = \{U\} \qquad (1.4)$$

In structures it is usually easier to define boundaries in terms of displacements than applied 'loading' and thus the formulation of (1.3) is predominantly used. This method is called the Stiffness Method and involves the finding of a stiffness matrix. The complementary method is the Flexibility Method. For pin-jointed or beam structures, the stiffness method is ideal. However for the continuum, the stiffness method, though predominantly used, has some disadvantages due to the assumptions made about the description of displacements across an element. This initial assumption and the subsequent use of energy methods to find the stiffness matrix are an application of the theorem of minimum potential energy which leads to an upper bound on element stresses. Conversely the flexibility method is an example of the theorem of minimum complementary energy which leads to a lower bound on element stresses. The compromise led the development of the hybrid element which involves an averaging process.

The use of the flexibility method and the development of hybrid elements is beyond the scope of this text. The methods described herein and the programs presented herewith use the stiffness approach.

The elements used are all for linear elastic materials and small displacement theory is assumed throughout. No derivations of the basic relations are given since these are readily available in standard texts, several of which are suggested at the end of the chapter. Little knowledge of matrix manipulation is required beyond being able to multiply two matrices and recognize symmetry. The main matrix inversion and equation solution routines are standard and can be found in texts suggested at the end of the chapter. It is very useful to be able to think of sets of equations being expressed in matrix form and being able to manipulate them in this form, but (where possible) in this text the matrices have been multiplied out.

The use of energy methods is more complex and beyond the scope of this book. Essentially energy methods consist of the Principle of Virtual Work and the dual theorems of Minimum Potential and Minimum Complementary Energy, which can be applied to beam and continuum structures alike. Once again, a text is suggested at the end of the chapter.

The programs

The four programs **PJFRAME**, **PLFRAME**, **FEP** and **FEPB** do not form a suite of programs but there are some similarities. Possibly of more interest are the differences, which exhibit different approaches to implementing the finite element method. An overriding principle has been to keep the programs simple and the listings as readable as possible. A firm grasp of the philosophy demonstrated in these programs will enable the user to make use of larger packages and have a better understanding of the workings!

The framework programs have been restricted to two dimensions – plane frames, but an understanding of their workings should enable someone with programming experience to extend them to full three-dimensional space frames. The continuum element, the constant strain triangle or CST, is the simplest approximation, but stands as a good starting point to understanding higher order elements such as linear strain triangles, quadrilaterals and even isoparametric elements. The complexities of finite elements can be quite severe and some references are given at the end of the chapter to assist in any further study. The plate element presented is not the simplest of elements but will still not deal with the shear effect on plate deformation.

The data is input in different ways. In **PJFRAME**, an interrogative or interactive approach is adopted, where the data is requested on the micro-computer screen and typed in thereon. There are prompts for each nodal point and each element and several other requests are answered by YES or NO. The data is then held in store while the program is active, but once the computer is switched off the data disappears. In addition, should something go wrong with the typing in of data, sometimes the data already typed in might be lost. Thus this method of data input is used only in **PJFRAME**. In all other programs the data is set up in DATA statements at the end of the program and can thus be saved on tape or disc with the program. In all cases, as is good practice, all input data is printed out on a printer and the user is asked to check this before letting the program proceed further.

When it comes to evaluating the stiffness matrices of the structure, two approaches are used. In **PJFRAME**, the complete uncondensed structure stiffness matrix is evaluated and then condensed by progressively incorporating the displacement boundary condition until there are exactly the number of equations corresponding to the number of unknown displacements. In the other three programs, knowledge of the boundary condition is used when transferring elements stiffness coefficients into the structure stiffness matrix, which is thus produced in its condensed form. This latter approach can be much more efficient in terms of storage since the largest matrix or array in the program is the structure stiffness matrix. It should be noted that although the stiffness matrices are symmetrical, and only half need be stored, all programs store the complete square matrix for ease of programming – this is an extravagance. Also, where a structure contains many elements, the stiffness matrix can become sparse (with many zero terms) and with all the terms clustered round the diagonal, or in other words 'banded'. Such banding can lead to another far more efficient way of storing the stiffness matrix but the programming complexities build up and clarity drops. Thus the simple uncluttered (but inefficient) storing of the complete stiffness matrix is used here. Comments on more efficient storage can be found in Appendix A.2.

The solution procedure for solving the resulting linear algebraic simultaneous equations is identical in all four programs – Gaussian Elimination. Once again this is very simply presented and makes no use of the symmetry and possible banding of the stiffness matrix of coefficients. It makes no attempt to solve the equations in a way which will reduce numerical inaccuracies. There are nowadays some very elegant ways of solving the equations from structural problems, especially 'frontal solvers', but the programming complexities are formidable.

The determining of the unknown nodal displacements is always followed by a back substitution procedure to determine element or member forces, stresses or moments depending on the program. This is usually done by assembling various element matrices together and multiplying these by the vector of the element's own displacement vector. This process is unique to each program.

The final part of each program is common for all although its implementation is different for the plate program **FEPB**. If the complete uncondensed stiffness matrix is built up for any structure and multiplied by the complete nodal displacement vector, a complete set of nodal forces results. These equations include the zero displacement boundary conditions of nodes where reaction forces and moments exist and so the complete displacement vector will, in fact, include several zeros. (The equations corresponding to these zeros are the redundant equations which are eliminated to produce the condensed set of equations and the condensed stiffness matrix). However, the complete set of nodal forces finally produced include three main groups of forces, which should be checked.

(1)   At nodes where known loads are applied, the forces should equal these

applied loads to within a certain degree of accuracy – this is a good check on the solution.

(2) The forces at nodes, which are not boundaries and which do not sustain applied loads, should be zero – once again this is a good check.

(3) The forces at nodes where zero boundary conditions have been imposed are the reactions – reaction forces for zero displacement nodes and reaction moments at zero slope or rotation nodes.

The procedure to determine nodal forces can be done in two ways. In the first three programs, the full uncondensed structure matrix is built up and multiplied by the full uncondensed displacement vector. However the storing of the uncondensed matrix is very 'expensive' on storage, especially if no account is taken of banding or symmetry. The approach taken in the last program is different. If an element stiffness matrix is multiplied by its element displacement vector, the element's nodal forces are found. If any other elements which share a node in common with this first element are similarly dealt with, a set of element nodal forces are found which operate at this common structure node and the algebraic sum of these forces gives the structure's nodal force. If the node is not sustaining an external force, the sum is zero, or if the node is sustaining an external load or is a boundary, the non-zero value of the sum is equal to the applied load or the reaction force. Thus in the program **FEPB**, as each element is processed, a running total of element nodal forces is kept, and the final result is a full set of nodal forces, and the requirement of storing the full uncondensed stiffness matrix is removed.

A final word about the programs and the book. They are not meant as expertly programmed and optimized structural analysis routines. Nor are the notes meant to be thorough and exhaustive. The hope is that the reader will, from the notes and the programs, see the philosophy behind the finite element method, appreciate the beauty of the approach, and subsequently either use larger programs and packages with more understanding and confidence or wish to seek further details of the method. No matter what, the book sets out to be a useful primer.

Appendix 1.1: Suggested further reading

Basic equations

    Baxter-Brown, J.McD. *Introductory Solid Mechanics,* Wiley, 1973.

    Benham, P. & Warnock, F.V. *Mechanics of Solids & Structures,* Pitman, 1981.

    Candall, S.H., Dahl, N.L. & Lardner, T.J. *Introduction to the Mechanics of Solids,* McGraw Hill, 1978.

    Timoshenko, S.P. & Goodier, J.N. *Theory of Elasticity,* McGraw Hill, 1970.

    Timoshenko, S.P. & Woinowsky Krieger, S. *Theory of Plates and Shells,* McGraw Hill, 1959.

Matrix Methods

    Jeffrey, A. *Mathematics for Engineers and Scientists,* Nelson, 1979.

Computer Routines

    Poole, L. & Borchers, M. *Some Common Basic Programs,* Osborne.

Energy Methods

    Richards, T.H. *Energy Methods in Stress Analysis,* Wiley, 1977.

Finite Element Analysis

    Desai, C.S. & Abel, J.F. *An Introduction to the Finite Element Method,* Van Nostrand Reinhold, 1972.

    Hinton, E. & Owen, D.R.J. *An Introduction to Finite Element Computations,* Pineridge, 1979.

    Coates, R.C., Coutie, M.G. & Kong, F.K. *Structural Analysis,* Nelson, 1980.

    Beaufait, F.W. *et al. Computer Methods of Structural Analysis,* Prentice Hall, 1970.

2 PJFRAME: Pin-jointed plane frames

2.0 Introduction

A plane structure can be defined as a number of members, joined together in a single plane with the ability to support loads, forces and moments.

A structure comprising members which are pin-jointed at their ends is called a truss. The examples dealt with here are two-dimensional or plane trusses, although the theory developed here can be extended to three-dimensional structures. The important simplifying assumption in trusses is that the pin joints are frictionless and thus can transmit no moment from one member to any other. Members can only sustain tension or compression, such members being called ties or struts respectively.

In practice the joints of a structure are made by welding, riveting or bolting. However, a simplified model with pin joints gives surprisingly good agreement with practical conditions. The more complex rigid jointed structures are dealt with in the next chapter.

The *sign convention* used will be as in Fig. 2.1.

compression (pushing on the pin)   tension (pulling on the pin)

Figure 2.1

Diplacements of nodes will be given the symbols u and v in the x- and y-directions respectively.

In order to illustrate the development of the computer solution method using pin-jointed elements, the more traditional and laborious approach is followed. The independent equations which form the basis of the stiffness approach to the solution are extracted from the resulting equations.

2.1 Example of truss

In order to illustrate the usual longhand solution procedure of pin-jointed structures, a relatively simple structure will be used.

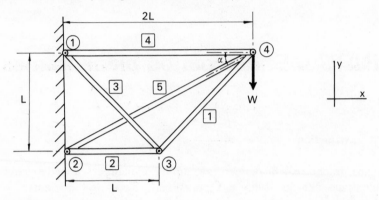

Figure 2.2

Consider the truss which is shown in Fig. 2.2, and which comprises 5 members and 4 joints or nodes. Joints ① and ② are firmly located in a rigid body and a load W is applied at Node ④.

Member	Length
[1]	$\sqrt{2}$L
[2]	L
[3]	$\sqrt{2}$L
[4]	2L
[5]	$\sqrt{5}$L

The Boundary Conditions are
u = v = 0 at ① and ②
Load W at ④

The angle α is such that $\cos\alpha = 2/\sqrt{5}$ and $\sin\alpha = 1/\sqrt{5}$

The solution is in three stages: (i) equilibrium of forces
(ii) compatibility of deformations
(iii) material properties.

2.1.1 Application of equilibrium

Initially assume that all members are in tension (Fig. 2.3) and consider equilibrium of each node in turn. T_1, T_2 etc. are the tensions in the members and H, V are the reaction forces at the supports.

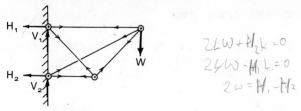

handwritten:
$2LW + H_2 k = 0$
$2kW - H_1 L = 0$
$2\omega = H_1 - H_2$

Figure 2.3

At Node ④, Fig. 2.4, resolving forces horizontally and vertically leads to

$$- T_4 - T_5 \cos\alpha - T_1 \cos 45° = 0$$
$$- T_1 \cos 45° - T_5 \sin\alpha - W = 0$$

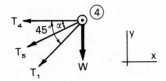

Figure 2.4

Substituting for $\cos\alpha$ and $\sin\alpha$ gives

$$T_4 + T_5\, 2/\sqrt{5} + T_1/\sqrt{2} = 0 \tag{2.1}$$

$$T_1/\sqrt{2} + T_5/\sqrt{5} + W = 0 \tag{2.2}$$

At Node ③, Fig. 2.5, equilibrium gives

$$-T_3/\sqrt{2} - T_2 + T_1/\sqrt{2} = 0 \tag{2.3}$$

$$1/\sqrt{2}\ T_1 + 1/\sqrt{2}\ T_3 = 0 \tag{2.4}$$

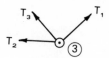

Figure 2.5

At Node ①, Fig. 2.6,

$$T_2 + T_5\, 2/\sqrt{5} - H_1 = 0 \tag{2.5}$$

$$V_1 + T_5\, 1/\sqrt{5} = 0 \tag{2.6}$$

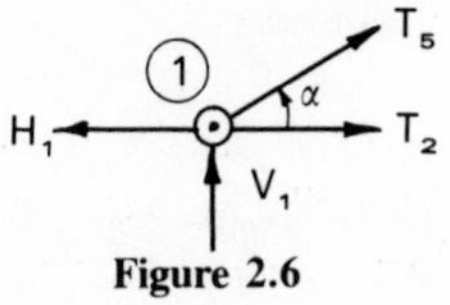

Figure 2.6

At Node ②, Fig. 2.7,

$$T_4 + T_3 / \sqrt{2} - H_2 = 0 \qquad (2.7)$$

$$V_2 - T_3 / \sqrt{2} = 0 \qquad (2.8)$$

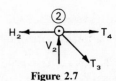

Figure 2.7

However the *eight* equations (2.1) to (2.8) have *nine* unknowns, T_1 to T_5, H_1, H_2, V_1, V_2.

Therefore the solution is not possible by using only equilibrium, and consideration must be given to the geometry of deformation, *compatibility.* [The overall equilibrium equations which would give $H_1 = -2W$, $H_2 = 2W$ and $V_1 + V_2 = W$, are not independent but implicit in the equations (2.1) to (2.8).] The structure in Fig. 2.2 is thus *statically indeterminate.*

Statical Determinacy is a condition for which *all internal forces may be obtained by equilibrium alone.* The internal forces are only a function of applied loads and the structure's geometry.

In general three classes of structures can be defined (Fig. 2.8): (a) non-rigid, (b) just rigid, (c) over-rigid or redundant.

(a)   non-rigid: it is unstable any displacement from its equilibrium position will cause collapse; it is not a structure but a mechanism.

(b)   just rigid: removal of one member destroys rigidity and allows collapse of part or whole of the structure.

(c)   over-rigid: removal of one member does not destroy rigidity.

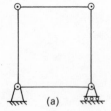
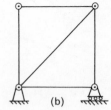
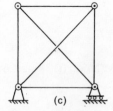

Figure 2.8

Consider the plane frames shown in Fig. 2.8.

(A) For each joint equilibrium requires $\Sigma F_x = 0$; $\Sigma F_y = 0$;
(B) For whole frame equilibrium requires $\Sigma F_x = 0$; $\Sigma F_y = 0$; $\Sigma M = 0$;

where F is a force in either coordinate direction and M a couple or moment about any point in the plane.

Thus only 3 reactions can be found from the 3 equations under (B). Let number of joints be j including support points and number of members m. There are 2j equations and m+r unknown forces and reactions; r has a minimum value of 3 (see (B) above) to cover general loading, although it may not be necessary for particular loading.

In the case of a plane frame

if m+r < 2j then structure is *unstable* or a mechanism
if m+r = 2j then structure is *statically determinate*
if m+r > 2j then structure is *statically indeterminate*.

Above is a *necessary* but not a *sufficient* condition as some arrangements of members could still provide a rigid system.

2.1.2 *Conditions for geometric compatibility*

Conisider a typical member ij of a frame, Fig. 2.9, having coordinates (x_i, y_i) at end i and (x_j, y_j) at end j. If the displacements of the node i are (u_i, v_i), and of the node j (u_j, v_j) then it is required to find the elongation e_{ij} of the member assuming small displacements. L_{ij} is the original length of member.

The equation for e_{ij} is derived as follows.

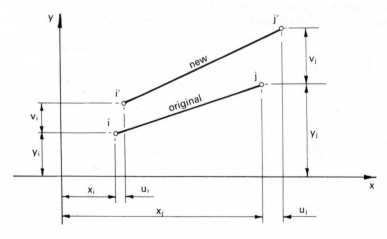

Figure 2.9

New length of member

$$= \left\{ \left[(x_j+u_j)-(x_i+u_i) \right]^2 + \left[(y_j+v_j)-(y_i+v_i) \right]^2 \right\}^{\frac{1}{2}}$$

$$= \left\{ \left[(x_j-x_i)+(u_j-u_i) \right]^2 + \left[(y_j-y_i)+(v_j-v_i) \right]^2 \right\}^{\frac{1}{2}}$$

$$= \left\{ (x_j-x_i)^2 + 2(x_j-x_i)(u_j-u_i)+(u_j-u_i)^2+(y_j-y_i)^2+2(y_j-y_i)(v_j-v_i)+(v_j-v_i)^2 \right\}^{\frac{1}{2}}$$

For small deformations the terms $(u_j-u_i)^2$ and $(v_j-v_i)^2$ may be neglected compared to the other terms and thus

$$\text{new length} \simeq \left\{ \left[(x_j-x_i)^2 + (y_j-y_i)^2 \right] + 2 \left[(x_j-x_i)(u_j-u_i)+(y_j-y_i)(v_j-v_i) \right] \right\}^{\frac{1}{2}}$$

and since $(x_j-x_i)^2 + (y_j-y_i)^2 = L_{ij}^2$ this leads to

$$\text{new length} = L_{ij} \left\{ 1 + 2 \left[\frac{(u_j-u_i)(x_j-x_i)+(v_j-v_i)(y_j-y_i)}{L_{ij}^2} \right] \right\}^{\frac{1}{2}}$$

The second term in the bracket $\{\ \}^{\frac{1}{2}}$ is small compared to unity and thus can be expanded by the binomial theorem to give

$$\text{new length} = L_{ij} \left\{ 1 + \tfrac{1}{2} 2 \left[\frac{(u_j-u_i)(x_j-x_i)}{L_{ij}^2} + \frac{(v_j-v_i)(y_j-y_i)}{L_{ij}^2} \right] \right\}$$

The change in length of the member e_{ij} is thus given

$$e_{ij} = \frac{(u_j-u_i)(x_j-x_i)}{L_{ij}} + \frac{(v_j-v_i)(y_j-y_i)}{L_{ij}} \tag{2.9}$$

The changes in lengths e_{ij} of all the members can now be expressed in terms of the displacements of the nodes; the e_{ij} constitute a compatible set of deformations.

Since the equation (2.9) is written in terms of the nodes i and j, at the end of each member, an origin of coordinates is required. Also members are required to be designated ij and so it is convenient to re-designate the forces and member lengths as shown in Fig. 2.10 and Tables 2.1 and 2.2.

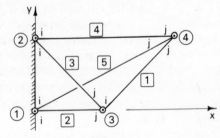

Figure 2.10

Table 2.1

Member	i	j	Length l_{ij}	Force
1	3	4	$l_1 = l_{34} = \sqrt{2}L$	$T_1 = T_{34}$
2	1	3	$l_2 = l_{13} = L$	$T_2 = T_{13}$
3	2	3	$l_3 = l_{23} = \sqrt{2}L$	$T_3 = T_{23}$
4	2	4	$l_4 = l_{24} = 2L$	$T_4 = T_{24}$
5	1	4	$l_5 = l_{14} = \sqrt{5}L$	$T_5 = T_{14}$

Table 2.2

Node	x	y
①	0	0
②	0	L
③	L	0
④	2L	L

Applying equation (2.9) to each member in turn leads to

$$\frac{T_{34}\sqrt{2}\,L}{EA}$$

$$
\begin{aligned}
e_{34} &= (u_4 - u_3)\,(2-1)\,/\sqrt{2} + (v_4 - v_3)\,(1-0)\,/\sqrt{2} \\
e_{13} &= (u_3 - u_1)\,(1-0)\,/\sqrt{1} + (v_3 - v_1)\,(0-0)\,/1 \\
e_{23} &= (u_3 - u_2)\,(1-0)\,/\sqrt{2} + (v_3 - v_2)\,(0-1)\,/\sqrt{2} \\
e_{24} &= (u_4 - u_2)\,(2-0)\,/2 + (v_4 - v_2)\,(1-1)\,/2 \\
e_{14} &= (u_4 - u_1)\,(2-0)\,/\sqrt{5} + (v_4 - v_1)\,(1-0)\,/\sqrt{5}
\end{aligned}
\qquad (2.10)
$$

With these five extra equations, however, 12 more unknowns have been added and before any solution is possible consideration must also be given to the properties of the structural material.

2.1.3 Stress/strain or force/elongation relations

Neither the derivation of the equations of equilibrium (2.1) to (2.8) nor the equations of compatibility (2.10) required the defining of a material – indeed the material in the truss could be steel, aluminium or plastic. However, in order to combine the resulting equations, the material properties must be considered. Under elastic conditions, the modulus of elasticity, E, is the only material property required, because truss members are in either simple tension or compression and *uniaxial stress-strain conditions apply.*

Strain $\epsilon = \sigma/E$ where σ is stress.

Strain can be defined here as extension/original length and stress as force/ area.

Thus for any member ij

$$\epsilon_{ij} = \frac{e_{ij}}{L_{ij}} = \frac{T_{ij}}{A_{ij}} \frac{1}{E_{ij}}$$

where A_{ij} is the member cross-section area.

This leads to

$$e_{ij} = \frac{T_{ij} \; L_{ij}}{(EA)_{ij}} \tag{2.11}$$

For the truss in Fig. 2.2 all members are assumed to have the same values of E and A.

Equation (2.11) can now be applied to each member.

$$\left.
\begin{array}{ll}
e_{34} = T_{34} \; \sqrt{2}L/EA & e_{24} = T_{24} \; 2L/EA \\[2mm]
e_{13} = T_{13} \; L/EA & e_{14} = T_{14} \; \sqrt{5}L/EA \\[2mm]
e_{23} = T_{23} \; \sqrt{2}L/EA &
\end{array}
\right\} \tag{2.12}$$

Five more equations have been found and so there are now 18 equations for 22 unknowns.

2.1.4 Boundary conditions and solution

The discrepancy of 4 equations is made up by adding the boundary conditions

$$u_1 = v_1 = u_2 = v_2 = 0$$

All 18 equations are written in their correctly subscripted form.

$$\left.
\begin{array}{ll}
(2.5) & T_{13} + T_{14}2/\sqrt{5} - H_1 = 0 \\[2mm]
(2.6) & V_1 + T_{14}/\sqrt{5} = 0 \\[2mm]
(2.7) & T_{24} + T_{23}/\sqrt{2} - H_2 = 0 \\[2mm]
(2.8) & V_2 - T_{23}/\sqrt{2} = 0 \\[2mm]
(2.3) & -T_{23}/\sqrt{2} - T_{13} + T_{34}/\sqrt{2} = 0 \\[2mm]
(2.4) & T_{34}/\sqrt{2} + T_{23}/\sqrt{2} = 0 \\[2mm]
(2.1) & -T_{24} - T_{14}\,2/\sqrt{5} - T_{34}/\sqrt{2} = 0 \\[2mm]
(2.2) & -T_{34} - T_{14}/\sqrt{5} - W = 0
\end{array}
\right\} \tag{2.13}$$

$$e_{34} = (u_4 - u_3)\,1/\sqrt{2} + (v_4 - v_3)\,1/\sqrt{2}$$

$$e_{13} = (u_3)\,1$$

$$e_{23} = u_3\;1/\sqrt{2} + v_3\,(-1/\sqrt{2}) \qquad (2.10)$$

$$e_{24} = u_4\,1$$

$$e_{14} = u_4\;2/\sqrt{5} + 2/\sqrt{5}$$

and

$$e_{34} = \sqrt{2}\;T_{34}/\alpha \qquad e_{24} = 2.T_{24}/\alpha$$

$$e_{13} = T_{13}/\alpha \qquad\qquad e_{14} = \sqrt{5}\;T_{14}/\alpha \qquad (2.12)$$

$$e_{23} = \sqrt{2}\;T_{23}/\alpha$$

substituting α for EA/L.

Equating equations (2.10) and (2.12) eliminates e_{ij}'s and gives relations between T_{ij} and u's, v's.

Therefore,

$$e_{34} = \sqrt{2}\;T_{34}/\alpha = (u_4 - u_3)\,1/\sqrt{2} + (v_4 - v_3)\,1/\sqrt{2}$$

$$T_{13}/\alpha = u_3$$

$$\sqrt{2}\;T_{23}/\alpha = u_3/\sqrt{2} - v_3/\sqrt{2}$$

$$2\,T_{24}/\alpha = u_4$$

$$\sqrt{5}\;T_{14}/\alpha = u_4\;2/\sqrt{5} + v_4\;1/\sqrt{5}$$

Substitution of the expressions for T_{ij} into equations (2.13) yields 8 equations in the 8 unknowns u_3, v_3, u_4, v_4, H_1, V_1, H_2, V_2 (Table 2.3).

Table 2.3 (2.14)

u_3		$+\dfrac{4}{5\sqrt{5}}\,u_4$	$+\dfrac{2}{5\sqrt{5}}\,v_4$	$-H_1/\alpha$	$=$	0
		$\dfrac{2}{5\sqrt{5}}\,u_4$	$+\dfrac{4}{5\sqrt{5}}\,v_4$	$+V_1/\alpha$	$=$	0
$\dfrac{1}{2\sqrt{2}}\,u_3$	$-\dfrac{1}{2\sqrt{2}}\,v_3$	$+\dfrac{1}{2}\,u_4$		$-H_2/\alpha$	$=$	0
$-\dfrac{1}{2\sqrt{2}}\,u_3$	$+\dfrac{1}{2\sqrt{2}}\,v_3$			$+V_2/\alpha$	$=$	0
$\left(-\dfrac{1}{\sqrt{2}}-1\right)u_3$		$+\left(\dfrac{1}{2\sqrt{2}}\right)u_4$	$+\left(\dfrac{1}{2\sqrt{2}}\right)v_4$		$=$	0
	$-\dfrac{1}{\sqrt{2}}\,v_3$	$+\dfrac{1}{2\sqrt{2}}\,u_4$	$+\dfrac{1}{2\sqrt{2}}\,v_4$		$=$	0
$\dfrac{1}{2\sqrt{2}}\,u_3$	$+\dfrac{1}{2\sqrt{2}}\,v_3$	$+\left(-\dfrac{1}{2}-\dfrac{4}{5\sqrt{5}}-\dfrac{1}{2\sqrt{2}}\right)u_4$	$+\left(-\dfrac{2}{5\sqrt{5}}-\dfrac{1}{2\sqrt{2}}\right)v_4$		$=$	0
$\dfrac{1}{2\sqrt{2}}\,u_3$	$+\dfrac{1}{2\sqrt{2}}\,v_3$	$+\left(-\dfrac{1}{2\sqrt{2}}-\dfrac{2}{5\sqrt{5}}\right)u_4$	$+\left(-\dfrac{1}{2\sqrt{2}}-\dfrac{1}{5\sqrt{5}}\right)v_4$		$=$	W/α

These eight linear algebraic simultaneous equations can be solved to give

$u_3 = -1.3299 \ \text{WL}/AE \qquad\qquad v_3 = -3.2107 \ \text{WL}/AE$

$u_4 = -2.6700 \ \text{WL}/AE \qquad\qquad v_4 = -9.0904 \ \text{WL}/AE$

$H_1 = -2W \qquad\qquad\qquad\quad\ \ V_1 = 0.3352 \ W$

$H_2 = 2W \qquad\qquad\qquad\quad\ \ \ V_2 = 0.6648 \ W$

However careful examination of equations (2.14) reveals that the last four equations are in the four unknowns u_3, v_3, u_4, v_4 and so the 8 equations can be decoupled.

Solution of the last 4 equations can be done separately and the remaining unknowns H_1, V_1, H_2, V_2 can then be evaluated separately by substituting u_3, etc., into the appropriate equation.

This general characteristic of the equations led to the beam element analysis of structures using the stiffness method. In this method, the structure is considered to be an assemblage of members and for each member there exists a unique force/displacement relationship, derived from general relations which incorporate compatibility and the stress/strain relations. The members are collected together using nodal equilibrium, and the resulting equations relate unknown nodal displacements to applied loads.

The general relations are derived in the following section.

2.2 The pin-jointed element

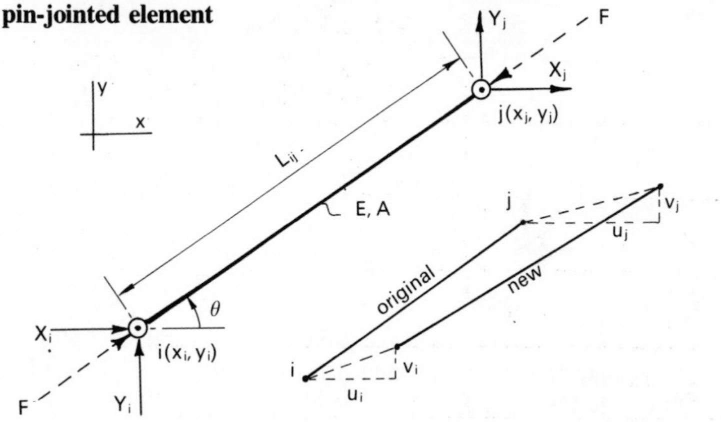

Figure 2.11

Consider the pin-jointed member ij subjected to a compressive force F, which has components X_i, Y_i and X_j, Y_j at the ends i and j. As a result of the structure deforming and of the member itself deforming, the ends i and j are displaced u_i, v_i and u_j, v_j. The member is originally at an angle θ to the positive x- direction.

From equation (2.9), the elongation is given by

$$e_{ij} = (u_j - u_i)\ \frac{(x_j - x_i)}{L_{ij}} + (v_j - v_i)\ \frac{(y_j - y_i)}{L_{ij}}$$

where L_{ij} is the length of the member.

Note that $(x_j - x_i)/L_{ij} = \cos\theta$ and $(y_j - y_i)/L_{ij} = \cos(90 - \theta) = \sin\theta$ and that these are usually given the symbols l, m and called direction cosines.

Thus

$$e_{ij} = \frac{-F.L}{AE} = \left[\frac{-X_i}{\cos\theta} \right] \frac{L}{AE} = \frac{-X_i}{l} \cdot \frac{L}{AE} \quad \text{since } F = \frac{X_i}{\cos\theta}$$

$$\therefore \quad X_i = \frac{-EA}{L}\, l.e_{ij} = \frac{-EA}{L} \left[(u_j - u_i)\, l^2 + lm(v_j - v_i) \right]$$

$$= -X_j \text{ from equilibrium}$$

Similarly, since $F = Y_i / \sin\theta = -Y_j / \sin\theta$

$$Y_i = \frac{EA}{L} \left[(u_i - u_j)\, lm + m^2(v_i - v_j) \right] = -Y_i$$

Finally

$$X_i = \frac{EA}{L} \left[(u_i-u_j)\, l^2 + lm\, (v_i-v_j) \right] = -X_j$$

$$Y_i = \frac{EA}{L} \left[(u_i-u_j)lm + m^2(v_i-v_j) \right] = -Y_j$$

(2.15)

Note: X and Y are the applied forces to the member ij at i and could be notated X_{ij} and Y_{ij} to distinguish them from structure nodal forces. Similarly, X_j and Y_j could be noted X_{ji} and Y_{ji}.

These can be expanded and written as

$$EA/L \left[l^2u_i + lmv_i - l^2u_j - lmv_j \right] = X_i \quad \Big\rbrace$$

$$EA/L \left[lmu_i + m^2v_i - lmu_j - m^2v_j \right] = Y_i$$

$$EA/L \left[- l^2u_i - lmv_i + l^2u_j + lmv_j \right] = X_j$$

$$EA/L \left[-lmu_i - m^2v_i + lmu_j + m^2v_j \right] = Y_j$$

or more concisely in matrix form as

$$EA/L \begin{bmatrix} l^2 & lm & -l^2 & -lm \\ lm & m^2 & -lm & -m^2 \\ -l^2 & -lm & l^2 & lm \\ -lm & -m^2 & lm & m^2 \end{bmatrix} \begin{bmatrix} u_i \\ v_i \\ u_j \\ v_j \end{bmatrix} = \begin{bmatrix} X_i \\ Y_i \\ X_j \\ Y_j \end{bmatrix}$$

(2.15a)

$$or \qquad [K_{ij}] \, \{U\} = \{Q\}$$

(2.16)

where $[K_{ij}]$ (the stiffness matrix of member ij) is square and *symmetric*,

$\{U_{ij}\}$ is the displacement vector

$\{Q\}$ is the load vector.

$[\mathbf{K}_{ij}]$ is the member stiffness matrix written with respect to a global coordinate system, i.e. not a coordinate system local to the member.

Note: The forces X_i etc. are applied forces to the member at the corresponding node.

2.2.1 Direction Cosines

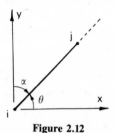

Figure 2.12

By definition, l and m are the direction cosines of the beam ij where

$$l = \cos\theta, m = \cos\alpha$$

$$= \cos(90\text{-}\theta)$$

$$= \sin\theta$$

The angle θ is defined as the *angle between the positive x- axis and the positive direction of the beam (defined as i to j)*

If $\theta = 45°$ say in Fig. 2.12 $l = 1/\sqrt{2}, m = 1/\sqrt{2}$

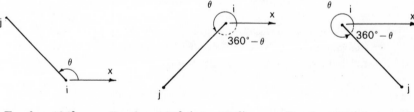

For $\theta = 135°$ For $\theta = 225°$ (or $-135°$)   For $\theta = 315°$ (or $-45°$)

$l = -1/\sqrt{2}$ $l = -1/\sqrt{2}$ $l = +1/\sqrt{2}$

$m = +1/\sqrt{2}$ $m = -1/\sqrt{2}$ $m = -1/\sqrt{2}$

The stiffness matrix $[\mathbf{K}_{ij}]$ is a matrix of sixteen stiffness coefficients which could be likened to spring stiffnesses. The coefficient on the diagonal relates the displacement u or v at a node to the applied force in the x- or y- direction at that same node. A force applied to one node in a structure displaces all the other unrestrained nodes and the ratio of the magnitude of this force to any nodal displacement can also be likened to a spring stiffness – these can be found as off-diagonal terms in the stiffness matrix $[\mathbf{K}_{ij}]$.

2.2.2 *Procedure in applying stiffness method to a pin-jointed structure*

To apply this method, it is useful to follow a strict procedure and avoid the derivation of redundant equations. This procedure is equivalent to deriving immediately the 'condensed' stiffness matrix (which will be discussed later in section 2.3.1).

(i)   Notate the structure with node numbers ①, ② etc. and member numbers ☐1, ☐2 etc.
Mark on i and j ends of the members.

(ii)   Tabulate the member properties.

Member	Nodes		θ	$l = \cos\theta$	$m = \sin\theta$	Length	If required	
	i	j					A	E

(iii)   Boundary conditions. Define zero displacement components and thus determine *unknown nodal displacements*.

(iv)   *Apply nodal equilibrium corresponding to unknown nodal displacements* viz. Y's for v's and X's for u's. Failure to do so will result in excess equations being derived.

(v)   Apply the element equations (2.15) which should result in a symmetrical stiffness matrix. Solve for the unknown nodal displacements.

(vi)   Determine required member forces.

(vii) Determine external reactions.

2.2.3 *Solution of truss in Fig. 2.10 using the stiffness method*

Follow the procedure outlined in 2.2.2.

(i)   Notate the structure – Fig. 2.13.

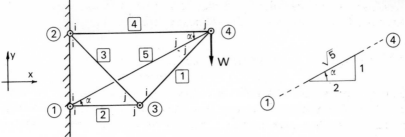

Figure 2.13

(ii)   Tabulate the member properties – Table 2.4.

Table 2.4

Member	Nodes i	Nodes j	θ	D. Cosines l	D. Cosines m	Length
①	3	4	45°	$1/\sqrt{2}$	$1/\sqrt{2}$	$\sqrt{2}\,L$
②	1	3	0°	1	0	L
③	2	3	– 45°	$1/\sqrt{2}$	$-1/\sqrt{2}$	$\sqrt{2}\,L$
④	2	4	0°	1	0	2L
⑤	1	4	$\cos^{-1} 2/\sqrt{5}$ $= \alpha$	$2/\sqrt{5}$	$1/\sqrt{5}$	$\sqrt{5}\,L$

(iii)   Define boundary conditions
$$u_1 = v_1 = u_2 = v_2 = 0$$
Determine unknown nodal displacements
$$u_3, v_3, u_4, v_4$$

(iv)   Apply nodal equilibrium corresponding to unknown nodal displacements. Since both u and v are unknown at nodes ③ and ④ equilibrium is applied in both coordinate directions at ③ and ④.

Node ③ – Fig. 2.14
$$X_3 = 0 = X_{31} + X_{32} + X_{34} \qquad (2.17)$$
$$Y_3 = 0 = Y_{31} + Y_{32} + Y_{34} \qquad (2.18)$$

X_3, Y_3 represent externally applied loads which are *sub-divided* among the member external (or applied) loads X_{31}, Y_{31} etc., and so equations (2.17),

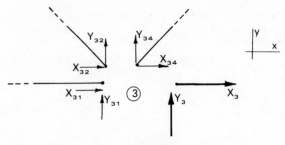

Figure 2.14

(2.18) above and (2.19), (2.20) below are not of the usual 'action-reaction' type. Here both X_3 and Y_3 are zero since the only load applied to the structure is at node ④.

Node ④ – Fig. 2.15

$$X_4 = 0 = X_{43} + X_{41} + X_{42} \tag{2.19}$$

$$Y_4 = -W = Y_{43} + Y_{41} + Y_{42} \tag{2.20}$$

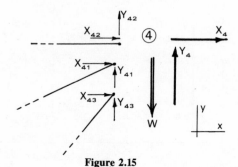

Figure 2.15

Note the negative sign in $Y_4 = -W$ since it is acting in a negative Y-direction. Application of equations (2.15) – (2.17) to (2.20) leads to

$$X_3 = 0 = \frac{-EA}{L} \left[(u_1 - u_3)\,1 + (u_1 - v_3)\,0 \right]$$

$$+ \quad \frac{-EA}{\sqrt{2}\,L} \left[(u_2 - u_3)\,\tfrac{1}{2} - \tfrac{1}{2}\,(u_2 - v_3) \right]$$

$$+ \left(\frac{EA}{\sqrt{2}L} \right) \left[(u_3 - u_4)\,\tfrac{1}{2} + \tfrac{1}{2}\,(v_3 - v_4) \right]$$

or

$$EA/L \left[u_3(1 + 1/\sqrt{2}) + v_3(0) + u_4(-1/2\sqrt{2}) + v_4(-1/2\sqrt{2}) \right] = 0 \tag{2.21}$$

$$Y_3 = 0 = -EA/L\ \big[(u_1 - u_3)\,0 + (v_1 - v_3)\,0\big]$$

$$+(-EA/\sqrt{2}L)\ \big[(u_2 - u_3)\,(-\tfrac{1}{2}) + \tfrac{1}{2}\,(v_2 - v_3)\big]$$

$$+\ ^{EA}/\sqrt{2}L\ \big[(u_3 - u_4)\,\tfrac{1}{2} + \tfrac{1}{2}\,(v_3 - v_4)\big]$$

or

$$EA/L\ \big[u_3\,0 + v_3\,{}^1\!/\sqrt{2} - 1/2\sqrt{2}\,u_4 - 1/2\sqrt{2}\,v_4\big] = 0$$

$$(2.22)$$

$$X_4 = 0 = -EA/\sqrt{2}L\ \big[(u_3 - u_4)\,\tfrac{1}{2} + (v_3 - v_4)\,\tfrac{1}{2}\big]$$

$$-EA/\sqrt{5}L\ \big[(u_1 - u_4)\,\frac{4}{5} + \frac{2}{5}\,(v_1 - v_4)\big]$$

$$-EA/2L\ \big[(u_2 - u_4)\,1 + 0\,(v_2 - v_4)\big]$$

or

$$EA/L\ \big[u_3(-1/2\sqrt{2}) + v_3\,(-1/2\sqrt{2}) + u_4\,(1/2\sqrt{2} + 4/5\sqrt{5} + \tfrac{1}{2})$$

$$+\ v_4\,(1/2\sqrt{2} + 2/5\sqrt{5})\big] = 0 \qquad (2.23)$$

$$Y_4 = -W = -\frac{EA}{\sqrt{2}L}\ \big[(u_3 - u_4)\,\tfrac{1}{2} + \tfrac{1}{2}\,(v_3 - v_4)\big]$$

$$-\frac{EA}{\sqrt{5}L}\ \big[(u_1 - u_4)\,\frac{2}{5} + \frac{1}{5}\,(v_1 - v_4)\big]$$

$$-\frac{EA}{2L}\ \big[(u_2 - u_4)\,0 + 0\,(v_2 - v_4)\big]$$

or

$$\frac{EA}{L}\ \Big[u_3\,(-\frac{1}{2\sqrt{2}}) + v_3\,(-\frac{1}{2\sqrt{2}}) + u_4\,(-\frac{1}{2\sqrt{2}} + \frac{2}{5\sqrt{5}})$$

$$+\ v_4\,(\frac{1}{2\sqrt{2}}) + \frac{1}{5\sqrt{5}})\Big] = -W \qquad (2.24)$$

Equations (2.21) – (2.24) are identical to the last 4 equations of (2.14) when multiplied through by a negative sign. They can be solved longhand as simultaneous equations or put into matrix form and solved by one of many standard routines.

$$\frac{EA}{L}\begin{bmatrix} (1+\frac{1}{\sqrt{2}}) & 0 & -\frac{1}{2\sqrt{2}} & -\frac{1}{2\sqrt{2}} \\ 0 & \frac{1}{\sqrt{2}} & -\frac{1}{2\sqrt{2}} & -\frac{1}{2\sqrt{2}} \\ -\frac{1}{2\sqrt{2}} & -\frac{1}{2\sqrt{2}} & (\frac{1}{2\sqrt{2}}+\frac{4}{5\sqrt{5}}+\frac{1}{2}) & (\frac{1}{2\sqrt{2}})+\frac{1}{5\sqrt{5}} \\ -\frac{1}{2\sqrt{2}} & -\frac{1}{2\sqrt{2}} & (\frac{1}{2\sqrt{2}}+\frac{2}{5\sqrt{5}}) & (\frac{1}{2\sqrt{2}})+\frac{2}{5\sqrt{5}} \end{bmatrix}\begin{bmatrix} u_3 \\ v_3 \\ u_4 \\ v_4 \end{bmatrix} = \begin{bmatrix} 0 \\ 0 \\ 0 \\ -W \end{bmatrix} = \begin{bmatrix} X_3 \\ Y_3 \\ X_4 \\ Y_4 \end{bmatrix}$$

$$(2.25)$$

$$\text{or } [K]\{U\} = \{P\}$$

where $[K]$ is the global stiffness matrix of the structure

$\{U\}$ is the vector of unknown displacements

$\{P\}$ is the nodal load vector of the structure.

Multiply both sides by $[K]^{-1}$ to give

$$\{U\} = [K]^{-1}\{P\} \quad or \quad \begin{bmatrix} u_3 \\ v_3 \\ u_4 \\ v_4 \end{bmatrix} = [K]^{-1} \begin{bmatrix} X_3 \\ Y_3 \\ X_4 \\ Y_4 \end{bmatrix}$$

$$(2.26)$$

Therefore once the inverse $[K]^{-1}$ has been found, the product $[K]^{-1}\{P\}$ will give the unknown nodal displacements.

Just as the inverse of spring stiffness is spring flexibility, the inverse $[K]^{-1}$ is known as the *flexibility matrix*. Although the solution of (2.25) do not explicitly require the determination of $[K]^{-1}$, the flexibility matrix is useful to have as will be shown in section 2.2.4 below. Standard routines exist for finding the inverse of a matrix and for the matrix $[K]$ in (2.25) it is

$$[\mathbf{K}]^{-1} = \frac{L}{EA} \begin{bmatrix} 0.8250 & 0.5775 & -0.1751 & 1.3299 \\ 0.5775 & 2.8084 & -0.4228 & 3.2107 \\ -0.1751 & -0.4228 & 1.8246 & -2.6700 \\ +1.3299 & 3.2107 & -2.6700 & 9.0904 \end{bmatrix}$$

As expected $[\mathbf{K}]^{-1}$ is square and symmetric.

(vi) *Member forces:* Find the coordinate end forces of the member, X_{ij} and Y_{ij} and resolve these into the direction of the member.

e.g. Member ④ – Fig. 2.16

$$X_{42} = -\frac{EA}{2L}\left[(u_2 - u_4)\,1 + 0\,(v_2 - v_4)\right] = \frac{EA}{2L}\,u_4 \qquad \left.\begin{array}{c} \\ \\ \end{array}\right\} \quad \text{(no need to resolve)}$$

$$Y_{42} = 0$$

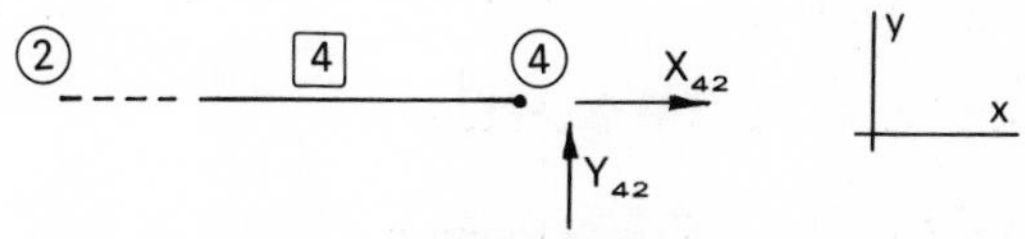

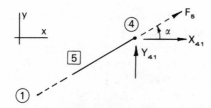

Figure 2.16

Member ⑤ – Fig. 2.17

$$X_{41} = -\frac{EA}{\sqrt{5}L}\left[(u_1 - u_4)\,\frac{4}{5} + \frac{2}{5}\,(v_1 - v_4)\right]$$

$$Y_{41} = -\frac{EA}{\sqrt{5}L}\left[(u_1 - u_4)\,\frac{2}{5} + \frac{1}{5}\,(v_1 - v_4)\right]$$

$$\therefore F_5 = X_{41}\cos\alpha + Y_{41}\sin\alpha$$

(Tension if positive, compression if negative)

Figure 2.17

(vii) *External reactions:* (Fig. 2.18)

At Node ① $X_1 = X_{14} + X_{13}$

$$= - \frac{EA}{\sqrt{5}L} \left[(u_1 - u_4) \frac{4}{5} + \frac{2}{5} (v_1 - v_4) \right]$$

$$+ \frac{EA}{L} \left[(u_1 - u_3) 1 + 0 (v_1 - v_3) \right]$$

$$= +2W$$

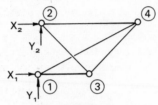

Figure 2.18

$$Y_1 = Y_{14} + Y_{13} = \frac{EA}{\sqrt{5}L} \left[(u_1 - u_4) \frac{2}{5} + \frac{1}{5} (v_1 - v_4) \right] + \frac{EA}{L} \left[(u_1 - u_3) 0 \right.$$

$$\left. + 0 (v_1 - v_3) \right]$$

$$= 0.3352 \ W$$

Similarly $X_2 = X_{23} + X_{24} = -2W$

and $Y_2 = Y_{23} + Y_{24} = 0.6648W$

2.2.4 *Load Vector* $\{P\}$

From equation (2.26) it can be seen that the unknown nodal displacements $\{U\}$ are found from the product $[K]^{-1} \{P\}$. It should be noted that the inverse is independent of the load vector and can be formed from knowledge of the geometry and material of the structure. The load vector $\{P\}$ for the particular loading shown in Fig. 2.13 is given by

$$\{P\} = \begin{bmatrix} 0 \\ 0 \\ 0 \\ -W \end{bmatrix}$$

The nodal displacements for the loading shown in Fig. 2.19 are found by multiplying the same inverse $[K]^{-1}$ (flexibility matrix) by a new load vector

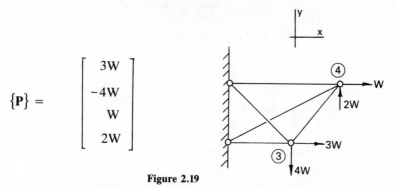

$$\{\mathbf{P}\} = \begin{bmatrix} 3W \\ -4W \\ W \\ 2W \end{bmatrix}$$

Figure 2.19

It should also be remembered that the flexibility matrix was symmetrical and this feature can be explained by reciprocity as follows:

Assume that a structure has a flexibility matrix $\begin{bmatrix} \mathbf{K} \end{bmatrix}^{-1}$ and the only load acting is a unit X-load at node i, then

$$\begin{bmatrix} u_i \\ \vdots \\ v_j \end{bmatrix} = \begin{bmatrix} \mathbf{K} \end{bmatrix}^{-1} \begin{bmatrix} 1 \\ 0 \\ \vdots \\ 0 \end{bmatrix}$$

Thus $v_j = \alpha$ for a unit X- load at i. Conversely if only a unit Y- load at j is applied

$$\begin{bmatrix} u_i \\ \vdots \\ v_j \end{bmatrix} = \begin{bmatrix} \mathbf{K} \end{bmatrix}^{-1} \begin{bmatrix} 0 \\ \vdots \\ 0 \\ 1 \end{bmatrix}$$

then $u_i = \alpha$ for a unit Y load at j.

2.2.5 *Nodal forces and reactions*

Reactions can be found by the method shown in section 2.2.3 (vii). However there is another method which not only finds the reactions but also checks the accuracy of the results. The method is laborious when done longhand but when the structure details are in the computer a few extra commands will accomplish the result. The steps are outlined below.

In section 2.2.3, only four equations were required to solve for the unknown displacements such that

$$
[K_c] \begin{bmatrix} u_3 \\ v_3 \\ u_4 \\ v_4 \end{bmatrix} = \begin{bmatrix} X_3 \\ Y_3 \\ X_4 \\ Y_4 \end{bmatrix}
$$

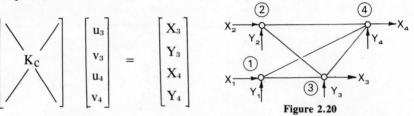

Figure 2.20

The 4 x 4 stiffness matrix $[K_c]$ is the condensed stiffness matrix which has eliminated the boundary conditions such as $u_1 = v_1 = u_2 = v_2 = 0$. However if equilibrium is applied to all nodes in the structure (Fig. 2.20) without the boundary conditions being incorporated then eight equations are obtained such that

$$
[K_u] \begin{bmatrix} u_1 \\ v_1 \\ u_2 \\ v_2 \\ u_2 \\ v_3 \\ u_4 \\ v_4 \end{bmatrix} = \begin{bmatrix} X_1 \\ Y_1 \\ X_2 \\ Y_2 \\ X_3 \\ Y_3 \\ X_4 \\ Y_4 \end{bmatrix} \qquad or \quad [K_u]\{U_u\} = \{P_u\}
$$

The 8 x 8 stiffness matrix $[K_u]$ is the uncondensed structure matrix.

Once all the unknown displacements are known, they can be combined wth the boundary conditions to form the uncondensed displacement vector. In the case of the simple structure shown in Fig. 2.13

$$
\{U_u\} = \begin{bmatrix} u_1 \\ v_1 \\ u_2 \\ v_2 \\ u_3 \\ v_3 \\ u_4 \\ v_4 \end{bmatrix} = \frac{WL}{AE} \begin{bmatrix} 0 \\ 0 \\ 0 \\ 0 \\ -1.3299 \\ -3.2107 \\ -2.6700 \\ -9.0904 \end{bmatrix}
$$

Thus the product $\begin{bmatrix} \mathbf{K}_u \end{bmatrix} \{\mathbf{U}_u\}$ will lead to

$$\{\mathbf{P}_u\} = \begin{bmatrix} X_1 \\ Y_1 \\ X_2 \\ Y_2 \\ X_3 \\ Y_3 \\ X_4 \\ Y_4 \end{bmatrix} = \begin{bmatrix} -2W \\ 0.3352W \\ 2W \\ 0.6648W \\ 0 \\ 0 \\ 0 \\ -W \end{bmatrix}$$

and of course X_1, Y_1, X_2, Y_2, are the boundary reaction forces in this structure. Additionally, the load vector $\{\mathbf{P}_u\}$ gives X_3, Y_3, X_4, as zero since no external loading is applied at nodes ③ and ④ in the X-direction or at node ③ in the Y-direction. The nodal force Y_4 will be $-W$. Not only does this method result in finding the reactions, but it also serves as a check on all the other externally applied nodal forces.

2.3 Implementation in program

2.3.1 *Description of program*

The program **PJFRAME** given in full in section 2.3.3 is written in an interactive mode such that data is inserted following prompts from the screen. The full description of the data input is given in section 2.3.2 below. Once all the data has been successfully inserted, the full analysis begins.

The program scans through each member and establishes the stiffness coefficients as found in equation (2.15a). In fact, since the coefficients in each quarter of $\begin{bmatrix} \mathbf{K}_{ij} \end{bmatrix}$ are identical in magnitude and with only a sign change, the program calculates the four coefficients l^2, lm, lm and m^2 in each quarter, multiples them by $+1$ or -1 and then adds them into the structure uncondensed stiffness matrix as shown below. The quarters are dealt with in the order shown

$$\begin{bmatrix} \mathbf{K}_{ij} \end{bmatrix} = \begin{bmatrix} \begin{matrix} x & & x \\ x & 2nd & x \end{matrix} & \begin{matrix} x & & x \\ x & 3rd & x \end{matrix} \\ \begin{matrix} x & & x \\ x & 1st & x \end{matrix} & \begin{matrix} x & & x \\ x & 4th & x \end{matrix} \end{bmatrix}$$

In order to locate the stiffness coefficient into the uncondensed stiffness matrix, one must have decided the form of that matrix beforehand.

The matrix with the displacement and load vectors for a structure with n nodes is of the form

$$
\begin{bmatrix}
k_{1,1} & k_{1,2} & k_{1,3}\dots\dots k_{1,2n} \\
k_{2,1} & k_{2,2}\dots\dots\dots\dots\dots\dots \\
k_{3,1}\dots\dots\dots\dots\dots\dots\dots\dots\dots \\
k_{4,1}\dots\dots\dots\dots\dots\dots\dots\dots\dots \\
k_{2(n-1),1}\dots\dots\dots\dots\dots\dots k_{2(n-1),2n} \\
k_{2n,1}\dots\dots\dots\dots k_{2n(n-1)}\quad k_{2n,2n}
\end{bmatrix}
\begin{bmatrix}
u_1 \\ v_1 \\ u_2 \\ v_2 \\ u_n \\ v_n
\end{bmatrix}
=
\begin{bmatrix}
X_1 \\ Y_1 \\ X_2 \\ Y_{i2} \\ X_n \\ Y_n
\end{bmatrix}
$$

This is identical to the method shown in section 2.2.5. Now one has the difficulty of locating particular stiffness coefficients from any member into the much larger structure stiffness matrix. To illustrate this consider a node ⑥, Fig. 2.21 in a structure to which are attached, say, three members ②, ③, ④ joining ⑥ to ⑤, ⑦, ⑧, and let there be (say) non-zero external forces applied to this structure of P_x, P_y

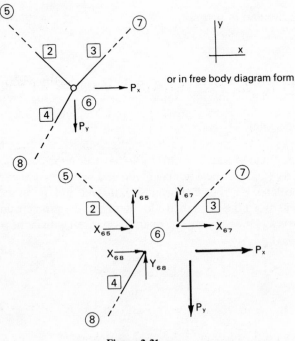

or in free body diagram form

Figure 2.21

Let member ② have $(\dfrac{EA}{L})_2$ and direction cosines l_2, m_2 and similarly for members ③ ④.

The two equilibrium equations for node ⑥ are

$$X_{65} + X_{67} + X_{68} = P_x \ (=X_6) \tag{2.27}$$

$$Y_{65} + Y_{67} + Y_{68} = -P_y \ (=Y_6) \tag{2.28}$$

The global uncondensed structure stiffness matrix has the form shown in Table 2.5.

All the coefficients from equations (2.27) and (2.28) will be on rows 11 and 12 respectively of the stiffness matrix. Expanding equation (2.27) leads to

$$(\frac{EA}{L})_2 \ \left[(u_6 - u_5)\, l_2^2 + l_2 m_2\, (v_6 - v_5) \right]$$

$$+ \ (\frac{EA}{L})_3 \ \left[(u_6 - u_7)\, l_3^2 + l_3 m_3\, (v_6 - v_7) \right]$$

$$+ \ (\frac{EA}{L})_4 \ \left[(u_6 - u_8)\, l_4^2 + l_4 m_4\, (v_6 - v_8) \right] = P_x$$

or

$$\left[-(\frac{EA}{L})_2\, l_2^2 \right] u_5 + \left[-(\frac{EA}{L})_2\, l_2 m_2 \right] v_5 + \left[(\frac{EA}{L})_2 l_2^2 + (\frac{EA}{L})_3\, l_3^2 \right.$$

$$+ \ (\frac{EA}{L})_4\, l_4^2 \right] u_6 + \left[(\frac{EA}{L})_2\, l_2 m_2 + (\frac{EA}{L})_3\, l_3 m_3 + (\frac{EA}{L})_4\, l_4 m_4 \right] v_6$$

$$+ \ \left[-(\frac{EA}{L})_3\, l_3^2 \right] u_7$$

$$+ \ \left[-(\frac{EA}{L})_3\, l_3 m_3 \right] v_7 + \left[-(\frac{EA}{L})_4\, l_4^2 \right] u_8 + \left[-(\frac{EA}{L})_4\, l_4 m_4 \right] v_8 = P_x$$

or $$\tag{2.29}$$

$$K_{11,9}\, u_5 + K_{11,10}\, v_5 + K_{11,11}\, u_6 + K_{11,12}\, v_6$$

$$+ \ K_{11,13}\, u_7 + K_{11,14}\, v_7 + K_{11,15}\, u_8 + K_{11,16}\, v_8 = P_x$$

The subscripts α, β of $K_{\alpha\beta}$ are calculated as follows:

(i)   when dealing with equilibrium at node i:
for the x-direction, $\alpha = (2i - 1)$
for the y-direction $\alpha = 2i$

Table 2.5

ROW	1	2	9	10	11	12	13	14	15	16
1										
2										
11			$K_{11,9}$	$K_{11,10}$	$K_{11,11}$	$K_{11,12}$	$K_{11,13}$	$K_{11,14}$	$K_{11,15}$	$K_{11,16}$
12			$K_{12,9}$	$K_{12,10}$	$K_{12,11}$	$K_{12,12}$	$K_{12,13}$	$K_{12,14}$	$K_{12,15}$	$K_{12,16}$
			u_5	v_5	u_6	v_6	u_7	v_7	u_8	v_8

$$
\begin{bmatrix} u_1 \\ v_1 \\ u_5 \\ v_5 \\ u_6 \\ v_6 \\ u_7 \\ v_7 \\ u_8 \\ v_8 \\ \vdots \end{bmatrix}
=
\begin{bmatrix} \vdots \\ P_x \\ -P_y \\ \vdots \end{bmatrix}
$$

(ii)   for member ij attached at node i, the β subscripts of the coeefficients of u_i, v_i, u_j, v_j are

$$\beta = (2i - 1),\ 2i,\ (2j - 1),\ 2j \text{ respectively.}$$

The coefficients $K_{11,11}$ and $K_{11,12}$ are made up of three distinct stiffness coefficients from members $\boxed{2}$, $\boxed{3}$ and $\boxed{4}$. In the program these coefficients K are initially zero and as each member is processed the coefficients accumulate.

In the above example $K_{11,11}$ and $K_{11,12}$ are complete once member $\boxed{4}$ has been processed. The other coefficients in (2.29) are incomplete depending on what other members might be attached to nodes $\boxed{6}$, $\boxed{7}$ and $\boxed{8}$.

Consider now the exact order in which the program approaches this redistribution of coefficients by considering member $\boxed{3}$ for example

$$\left[k_{ij}\right]_{\boxed{3}} = \left(\frac{EA}{L}\right)_3 \begin{bmatrix} l^2 & lm & -l^2 & -lm \\ lm & m^2 & -lm & -m^2 \\ -l^2 & -lm & l^2 & lm \\ -lm & -m^2 & lm & m^2 \end{bmatrix} \equiv \left[\begin{array}{cc|cc} k_{11} & k_{12} & k_{13} & k_{14} \\ k_{21} & k_{22} & k_{23} & k_{24} \\ \hline k_{31} & k_{32} & k_{33} & k_{34} \\ k_{41} & k_{42} & k_{43} & k_{44} \end{array}\right] .$$

$$\quad\quad u_6 \quad\quad v_6 \quad\quad u_7 \quad\quad v_7$$

k_{31} is added into $K_{13,11}$ then $k_{1,1}$ is added into $K_{11,11}$
k_{32} is added into $K_{13,12}$ $k_{1,2}$ is added into $K_{11,12}$
k_{41} is added into $K_{14,11}$ $k_{1,3}$ is added into $K_{12,11}$
k_{42} is added into $K_{14,12}$ $k_{1,4}$ is added into $K_{12,12}$

then $k_{1,3}$ is added into $K_{11,13}$ then $k_{3,3}$ is added into $K_{13,13}$
$k_{1,4}$ is added into $K_{11,14}$ $k_{3,4}$ is added into $K_{13,14}$
$k_{2,3}$ is added into $K_{12,13}$ $k_{4,3}$ is added into $K_{14,13}$
$k_{2,4}$ is added into $K_{12,14}$ $k_{4,4}$ is added into $K_{14,14}$

In the program the uncondensed matrix is stored for later use as $\left[\mathbf{KH}\right]$.

Condensing of stiffness matrix. The stiffness matrix as established above is the condensed matrix, which is not suitable for solving for the unknown nodal displacements, although it is useful for checking the solution by calculating nodal forces and reactions once all the nodal displacements are known (see section 2.2.5). The condensing of the matrix is the way in which the zero or known displacement boundary conditions are incorporated into the problem. Should a zero displacement exist at some boundary node, one row and the corresponding column are eliminated from the matrix, keeping the matrix still square and symmetrical.

If at node i, $u_i = 0$ then row $(2i-1)$ and column $(2i-1)$ are eliminated and so all subscripts of rows $> (2i-1)$ and columns $> (2i-1)$ reduce by one. If in addition $v_i = 0$ then the original row $2i$ and column $2i$ are eliminated, and so on.

In the program, each node i is labelled with a KODE which is either 1 or 2 or 3 or 4 corresponding to $u_i = v_i = 0$ or $v_i = 0$ or $u_i = 0$ or a free (unrestrained) node. The condensing procedure in the program considers each nodal displacement in turn, starting with u_1, and ending with v_n at the last node. If the u or v is zero the row and corresponding column are eliminated and the stiffness matrix progressively 'packed up' to the top left hand corner. The final condensed stiffness matrix is held in $[K]$ and is d x d in size where d is the number of unknown nodal displacements or *degrees of freedom*.

Load vector. The uncondensed load vector is progressively built up as the data is fed in node by node and so for an n-noded truss the load vector $\{PO\}$ is n x 1 in size. This like the stiffness matrix is condensed to a d x 1 matrix and held in $\{P\}$.

Solution procedure. The solution of the simultaneous equations uses a very simple Gaussian elimination technique which takes no account of the matrix symmetry or whether the matrix is heavily banded. More efficient methods exist and could be inserted. The coefficients of the stiffness matrix are held in $[K]$ and the load vector (the right-hand side of the equations), in $\{P\}$. After the solution is complete the result is held in the vector $\{P\}$.

As an example of how the elimination process works consider the following three simultaneous linear algebraic equations

$$x + 2y + 3z = 14$$
$$4x + 5y - 6z = -4$$
$$7x + 8y - 9z = -4$$

or

$$
\begin{matrix}
① \to \\
② \to \\
③ \to
\end{matrix}
\begin{bmatrix} 1 & 2 & 3 \\ 4 & 5 & -6 \\ 7 & 8 & -9 \end{bmatrix}
\begin{bmatrix} x \\ y \\ z \end{bmatrix}
=
\begin{bmatrix} 14 \\ -4 \\ -4 \end{bmatrix}
$$

(i) Eliminate x in two steps

$$
\begin{matrix}
② \times ① \Rightarrow \\
\\
= \dfrac{4}{1}
\end{matrix}
\begin{bmatrix} 1 & 2 & 3 \\ 0 & -3 & -18 \\ 7 & 8 & -9 \end{bmatrix}
\begin{bmatrix} x \\ y \\ z \end{bmatrix}
=
\begin{bmatrix} 14 \\ -60 \\ -4 \end{bmatrix}
$$

$$③ \times ① \Rightarrow \quad \begin{bmatrix} 1 & 2 & 3 \\ 0 & -3 & -18 \\ 7 & -6 & -30 \end{bmatrix} \begin{bmatrix} x \\ y \\ z \end{bmatrix} = \begin{bmatrix} 14 \\ -60 \\ -102 \end{bmatrix}$$
$$= \frac{7}{1}$$

(ii)   Eliminate y in one step

$$③ \times ② \Rightarrow \quad \begin{bmatrix} 1 & 2 & 3 \\ 0 & -3 & -18 \\ 0 & 0 & +6 \end{bmatrix} \begin{bmatrix} x \\ y \\ z \end{bmatrix} = \begin{bmatrix} 14 \\ -60 \\ 18 \end{bmatrix}$$
$$= \frac{-6}{-3}$$

(iii) Thus solve for z : $6z = 18 \; OR \; z = 3$

(iv) Back substitute $z = 3$ into ②
$$-3y - 18z = -60 \quad y = 2$$

(v) Back substitute $z = 3$, $y = 2$ into ①
$$x + 2y + 3z = 14 \quad x = 1$$

This method can obviously be extended to more than three equations. It is not suitable for ill-conditioned equations which might result for conditionally stable structures such as that shown in Fig. 2.22.

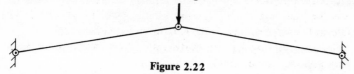

Figure 2.22

Member elongation, force and stress. For each member the elongation, force and stress are calculated. The solution gives the displacements u and v at all nodes in the structure, and to find the elongation of the member ij the net displacement of one end relative to the other must be found (see Fig. 2.11).

$$du_{ij} = u_j - u_i$$

$$dv_{ij} = v_j - v_i$$

and so for small displacements, elongation $e_{ij} = du_{ij} \cos \theta + dv_{ij} \sin \theta$

$$= l_{ij} \, du_{ij} + m_{ij} \, dv_{ij}$$

(Equation (2.9) could also be as such.)

The force in the member can thus be calculated from $F_{ij} = e_{ij} \left(\dfrac{EA}{L} \right)_{ij}$
and the stresses from $\sigma_{ij} = F_{ij} / A_{ij}$

Nodal forces. Finally the method described in section 2.2.5 is used to check the solution and to find reaction forces. The uncondensed stiffness matrix $[\mathbf{KH}]$ is multiplied by the uncondensed displacement vector, resulting in the co-ordinate external nodal forces X and Y at each node.

As a check:
- (i) nodes with no applied loads should have X = Y = 0
- (ii) loaded nodes should have X and/or Y equal to the given applied load.
- (iii) equilibrium requires that the sums of all X's and Y's should be zero.

2.3.2 *Data preparation*

A data preparation sheet is found in Appendix 2.1. The data is required in the following order and where more than one item is required for entry, a comma is used a separator. At several points in the program, questions are asked and should be answered YES or NO.

Structure data:
- (i) No. of members in the structure, NM.
- (ii) No. of modes in the structure, NN.

Note as written the program will solve for structures of up to 20 members (MX) and 10 nodes (NX = 10). To change these limits, MX and NX should be changed.

Nodal data. For each node –
- (iii) Node number – start with the first node as node ① and there should be no gaps.
- (iv) KODE   = 1 for $u_i = v_i = 0$, fixed node
            = 2 for $v_i = 0$, horizontal roller
            = 3 for $u_i = 0$, vertical roller
            = 4, free node
    (see Fig. A2.1)
- (v) Coordinates x, y of node
- (vi) Applied PX and PY forces at the node

Member data. For each member
- (vii) Member number, i and j end nodes, no matter which end of the member is given first. However, the members should start with number ☐1 and there should be no gaps.
- (viii) Member area A and modulus of elasticity E.

Notes. (a) The member numbers and nodal numbers should start at one and end with the last number with no gaps, although the data can be put in any member order or node order. Should any data be required to be changed only that member or node data need be input.

(b) Units are as decided by the user but must be consistent. For example, if E is in units of N/mm^2, then coordinates x and y are in mm and area A in mm^2 and forces PX, PY in N. The output is also in the corresponding units.

(c) It is recommended that member and nodal data be printed out and carefully checked before proceeding.

2.3.3   Program *PJFRAME*

Flow chart for **PJFRAME**

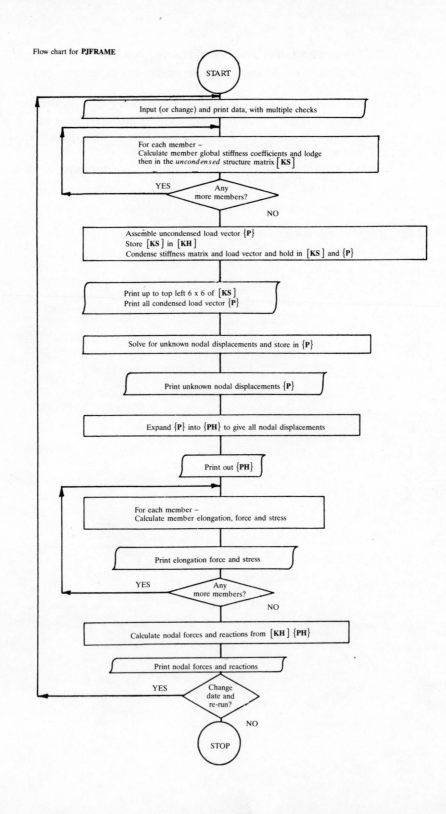

```
100 REM                 *** PROGRAM PJFRAME ***
105 REM               AUTHOR:
110 REM                   DAVID K. BROWN
112 REM          DEPARTMENT OF MECHANICAL ENGINEERING
115 REM                 UNIVERSITY OF GLASGOW
120 REM                      SCOTLAND
125 REM
130 REM                    AUGUST  1983
135 REM
140 REM    **   THE PROGRAM WILL ANALYSE PLANE
141 REM         FRAME/TRUSS PROBLEMS WITH UP TO
142 REM         10 NODES AND 20 MEMBERS.              **
145 REM
146 REM    **   LOADING IS APPLIED THROUGH COORDINATE
147 REM         NODAL FORCES AND 4 TYPES OF DISPLACEMENT
148 REM         BOUNDARY CONDITIONS ARE ALLOWED.      **
149 REM
150 REM    **   DATA INPUT IS FROM THE USER VIA
151 REM         THE KEYBOARD (IN RESPONSE TO PROMPTS
152 REM         APPEARING ON THE SCREEN).
153 REM
155 REM    **   OUTPUT CONSISTS OF:
156 REM                 NODAL DISPLACEMENTS AND FORCES;
157 REM                 MEMBER FORCES, STRESSES
158 REM                 AND ELONGATIONS.             **
160 REM
170 REM
200 PRINT"PLEASE WAIT WHILE ARRAYS ARE
210 PRINT"     BEING DIMENSIONED"
220 PJ$="PJFRAME"
225 GOSUB 53000  :REM MACHINE SPECIFIC STRINGS
230 MX=20  : REM MAX # OF MEMBERS
240 NX=10  : REM MAX # OF NODES
245 NF=2*NX :REM MAX # OF DEGREES OF FREEDOM
260 DIM NI(MX),NJ(MX),A(MX),E(MX)
270 DIM P(NF),PO(NF)
280 DIM X(NX),Y(NX),KODE(NX)
290 DIM KS(NF,NF),KH(NF,NF)
300 DIM PH(NF),F(NF),SP$(NX)
310 IR = 0
320 FORI=1TONF
330 PO(I)=0.
340 NEXTI
350 FORI=1TONF
360 P(I)=0.
370 PH(I) = 0.
380 F(I) = 0.
390 FOR J=1 TO NF
400 KH(I,J) = 0.
410 KS(I,J) = 0.
420 NEXT J
430 NEXT I
440 IF IR = 1 THEN 620
450 REM   **   DATA IS NOW REQUESTED           **
460 TM$="TOO MANY ":MA$=" MAXIMUM ALLOWED = "
470 PRINT"HOW MANY MEMBERS IN THE STRUCTURE   ";AK$;L4$;
475 INPUT NM
480 IFNM>MX THEN PRINT TM$;"MEMBERS,";MA$;MX:GOTO 470
```

```
490 PRINT "HOW MANY NODES   ";AK$;L4$;:INPUT NN
500 IFNN>NX THEN PRINT TM$;"NODES,";MA$;NX:GOTO 490
510 PRINT "NODAL DATA IS REQUIRED"
520 PRINT"     NODE NUMBER   ";AK$;L4$;:INPUT I
525 IF I<1 OR I>NN THEN 520
530 PRINT "TYPE IN KODE (SEE SHEET) ";AK$;L4$;
535 INPUT KODE(I)
540 IFKODE(I)<1 OR KODE(I)>4 THEN530
550 PRINT "X,Y COORDS OF NODE";I;"   ";AK$;L4$;
560 INPUT X(I),Y(I):PRINT
570 PRINT "APPLIED X,Y FORCES "
580 PRINT "    AT NODE";I;"    ";AK$;L4$;
590 INPUT PO(2*I-1),PO(2*I):PRINT
600 PRINT "ANY MORE NODES OR CORRECTIONS  [Y/N] ";Y$;L4$;
605 INPUT AN$
610 IF LEFT$(AN$,1)="Y" THEN 520
620 PRINT"DO YOU WISH A PRINT-OUT"
630 PRINT "  OF THE NODAL DATA  [Y/N] ";Y$;L4$;
635 INPUT AN$
640 IF LEFT$(AN$,1)="N"THEN 770
650 GOSUB 51000 :REM OPEN PRINTER CHANNEL
660 P$="       ****     PROGRAM "+PJ$+"    ****"
665 GOSUB 50000:IP=2:GOSUB 50020
670 P$=" N KODE    X         Y"
680 P$=P$+"        PX         PY":GOSUB50000
690 FW=12:NS=3
700 FOR I=1 TO NN:P$=STR$(I)+"  "+STR$(KODE(I))
710 XS = X(I)     :GOSUB20040:P$=P$+XS$
720 XS = Y(I)     :GOSUB20040:P$=P$+XS$
730 XS = PO(2*I-1):GOSUB20040:P$=P$+XS$
740 XS = PO(2*I)  :GOSUB20040:P$=P$+XS$
750 GOSUB 50000
760 NEXT I
765 GOSUB 52000 :REM CLOSE PRINTER CHANNEL
770 PRINT"DO YOU WISH TO CORRECT THE DATA [Y/N]   ";N$;L4$;
775 INPUT AN$
780 IF LEFT$(AN$,1)="Y" THEN 510
790 IF LEFT$(AN$,1)<> "N" THEN770
800 IF IR = 1 THEN 910
810 N = 2*NN
830 PRINT "MEMBER DATA IS REQUIRED"
840 PRINT:PRINT:PRINT"MEMBER #, & I,J"
850 PRINT"OF ITS END NODES  ";AK$;L4$;:INPUT L,NI(L),NJ(L)
860 PRINT"AREA, ELASTIC MODULUS"
870 PRINT"OF THE MEMBER  ";AK$;L4$;:INPUT A(L),E(L)
880 PRINT:PRINT "ANY MORE MEMBERS OR "
890 PRINT"DO YOU WISH TO CORRECT THE DATA [Y/N]   ";Y$;L4$;
895 INPUT AN$
900 IF LEFT$(AN$,1)="Y" THEN840
910 PRINT "DO YOU WISH A PRINT-OUT"
920 PRINT "  OF MEMBER DATA  [Y/N]  ";Y$;L4$;:INPUT AN$
930 IF LEFT$(AN$,1)="N"THEN 1030
940 GOSUB 51000 :REM OPEN PRINTER CHANNEL
950 GOSUB50010
960 P$=" M   NI   NJ      AREA            E":GOSUB50000
970 FOR L=1 TO NM
975 P$=STR$(L)+"  "+STR$(NI(L))+"  "+STR$(NJ(L))+"   "
980 XS = A(L):GOSUB20000:P$=P$+XS$
990 XS = E(L):GOSUB10000:P$=P$+XS$
1000 GOSUB50000
```

```
1010 NEXT L
1020 GOSUB 52000 :REM CLOSE PRINTER CHANNEL
1030 PRINT"DO YOU WISH TO CORRECT THE DATA [Y/N]   ";N$;L4$;
1035 INPUT AN$
1040 IF LEFT$(AN$,1)="Y" THEN 840
1050 IF LEFT$(AN$,1)<> "N" THEN 1030
1060 PRINT "THE PROGRAM IS NOW RUNNING"
1070 REM **   SCAN THROUGH ALL MEMBERS DETERMINING
1080 REM       ALL THEIR STIFFNESS COEFFICIENTS AND
1085 REM       LODGING THEM IN THE UNCONDENSED
1090 REM       GLOBAL STIFFNESS MATRIX [KS].         **
1095 REM
1100 L1 = 0
1110 L=0
1120 L=L+1
1130 II = NI(L)
1140 JJ = NJ(L)
1150 K = 2*NJ(L)
1160 M = 2*NI(L)
1170 GOTO 3140
1180 Z = -1.
1190 J = 1
1200 ON J GOTO 1300,1240,1270,1210,1360
1210 Z = 1.
1220 K = 2*NJ(L)
1230 GOTO 1300
1240 Z = 1.
1250 K = 2*NI(L)
1260 GOTO 1300
1270 Z =  -1.
1280 M = 2*NJ(L)
1290 REM **   LOCATE STIFFNESS COEFFICIENTS
1295 REM       INTO STIFFNESS MATRIX.              **
1300 KS(K-1,M-1) =KS(K-1,M-1)+Z*CL*CL*E(L)*A(L)/DL
1310 KS(K-1,M  ) =KS(K-1,M  )+Z*CL*CM*E(L)*A(L)/DL
1320 KS(K  ,M-1) =KS(K  ,M-1)+Z*CM*CL*E(L)*A(L)/DL
1330 KS(K  ,M  ) =KS(K  ,M  )+Z*CM*CM*E(L)*A(L)/DL
1340 J=J+1
1350 GOTO 1200
1360 IF L-NM >=0 GOTO 1400
1370 IF L-NM <0 GOTO 1120
1380 REM **   HOLD UNCONDENSED LOAD VECTOR [P] IN [PH],
1390 REM       HOLD UNCONDENSED STIFFNESS MATRIX [KS] IN [KH]. **
1400 FORI=1TON
1410 P(I)=PO(I)
1420 PH(I)=P(I)
1430 NEXTI
1440 FOR I=1 TO N
1450 FOR J=1 TO N
1460 KH(I,J) =KS(I,J)
1470 NEXT J
1480 NEXT I
1490 REM **   USING KODE AT EACH NODE,
1495 REM       PROGRESSIVELY CONDENSE THE
1500 REM       STIFFNESS MATRIX [KS] AND HOLD
1505 REM       FINALLY IN [KS]; SIMILARLY
1510 REM       CONDENSE [P] INTO [P].              **
1520 MZ = 1
1530 FOR IJ = 1 TO NN
1540 ON KODE(IJ) GOTO 1870,1730,1590,1560
```

```
1550 PRINT"KODE("IJ"> WRONG":GOTO3380
1560 MZ = MZ + 2
1570 GOTO 1970
1580 REM
1590 FOR I = MZ TO N
1600 P(I) = P(I+1)
1610 FOR J = 1TO N
1620 KS(I,J) =KS(I+1,J)
1630 NEXT J
1640 NEXT I
1650 FOR J = MZ TO N
1660 FOR I = 1TO N
1670 KS(I,J)=KS(I,J+1)
1680 NEXT I
1690 NEXT J
1700 MZ = MZ + 1
1710 GOTO 1970
1720 REM
1730 FOR I = MZ TO N
1740 P(I+1) = P(I+2)
1750 FOR J = 1 TO N
1760 KS(I+1,J) =KS(I+2,J)
1770 NEXT J
1780 NEXT I
1790 FOR J=MZ TO N
1800 FOR I = 1 TO N
1810 KS(I,J+1) =KS(I,J+2)
1820 NEXT I
1830 NEXT J
1840 MZ = MZ + 1
1850 GOTO 1970
1860 REM
1870 FOR I = MZ TO N
1880 P(I) = P(I+2)
1890 FOR J =1 TO N
1900 KS(I,J) =KS(I+2,J)
1910 NEXT J
1920 NEXT I
1930 FOR J = MZ TO N
1940 FOR I = 1 TO N
1950 KS(I,J) = KS(I,J+2)
1960 NEXT I,J
1970 NEXTIJ: REM 'FOR' AT 1250
1980 MZ= MZ-1
1990 M= MZ
2000 GOSUB 51000  :REM OPEN PRINTER CHANNEL
2010 GOSUB50010
2020 P$="":MQ=MZ
2025 IFMQ>6THENMQ=6:P$="TOP LEFT 6X6 CORNER OF THE "
2030 P$=P$+"CONDENSED STIFFNESS MATRIX":GOSUB50000
2040 FW=12:NS=4  :REM ** DON'T INCREASE FW!  **
2050 FOR I = 1TO MQ:P$=""
2060 FOR J = 1 TO MQ
2070 XS = KS(I,J)  :GOSUB10040:P$=P$+XS$
2080 NEXT J:GOSUB50000:GOSUB 50010
2090 NEXTI
2100 FW=12:NS=4
2110 P$="     CONDENSED LOAD VECTOR":GOSUB50000
2120 FOR I = 1 TO MZ
2130 XS = P(I):GOSUB20000:P$="        "+XS$:GOSUB50000
```

```
2140 NEXT I
2145 GOSUB 52000  :REM CLOSE PRINTER CHANNEL
2160 PRINT"DO YOU WISH THE PROGRAM"
2170 PRINT "  TO CONTINUE   [Y/N]  ";Y$;L4$;:INPUT AN$
2180 IF LEFT$(AN$,1)="N"THEN 3380
2190 REM   **   SOLUTION OF EQUATIONS NOW BEGINS
2200 REM      BY GAUSSIAN ELIMINATION.          **
2210 M1=M-1
2220 FOR I = 1 TO M1
2230 L = I+1
2240 FOR J = L TO M
2250 IFKS(J,I) = 0 GOTO 2300
2260 FOR K = L TO M
2270 KS(J,K) =KS(J,K)-KS(I,K)*KS(J,I)/KS(I,I)
2280 NEXT K
2290 P(J) = P(J)-P(I)*KS(J,I)/KS(I,I)
2300 NEXT J
2310 NEXT I
2320 P(M) = P(M)/KS(M,M)
2330 FOR I = 1 TO M1
2340 K = M-I
2350 L = K+1
2360 FOR J = L TO M
2370 P(K) = P(K)-P(J)*KS(K,J)
2380 NEXT J
2390 P(K) = P(K)/KS(K,K)
2400 NEXT I
2410 REM   **   THE VECTOR OF UNKNOWN DISPLACEMENTS
2415 REM       IS HELD IN [P].                **
2420 GOSUB 51000  :REM OPEN PRINTER CHANNEL
2430 GOSUB50010
2440 P$="       VECTOR OF UNKNOWN DISPLACEMENTS":GOSUB50000
2450 FW=12:NS=4
2460 FOR I = 1 TO MZ
2470 XS = P(I):GOSUB 10040:P$="        "+XS$:GOSUB50000
2480 NEXT I
2490 REM   **  EXPAND VECTOR [P] INTO [PH] BY
2500 REM      INCORPORATING BOUNDARY DISPLACEMENT VALUES   **
2510 MS = 0.0
2520 MA = 0
2530 FOR IJ = 1TO NN
2540 IF KODE(IJ) <> 4 GOTO 2600
2550 MS = MS+2
2560 MA = MA+2
2570 PH(MS-1) = P(MA-1)
2580 PH(MS) = P(MA)
2590 GOTO 2750
2600 IF KODE(IJ)<>3 GOTO 2660
2610 MS = MS+2
2620 MA=MA+1
2630 PH(MS-1)=0.
2640 PH(MS) = P(MA)
2650 GOTO 2750
2660 IF KODE(IJ) <> 2 GOTO 2720
2670 MS = MS+2
2680 MA = MA+1
2690 PH(MS-1)=P(MA)
2700 PH(MS)=0.
2710 GOTO 2750
2720 MS=MS+2
```

```
2730 PH(MS-1)=0.
2740 PH(MS) =0.
2750 NEXT IJ
2760 GOSUB50010
2770 P$="VECTOR OF ALL NODAL DISPLACEMENTS":GOSUB50000
2780 P$=" N         UX              VY":GOSUB50000
2790 FW=15:NS=4
2800 FOR I=1 TO NN
2810 XS = PH(2*I-1):GOSUB 10040:P$=XS$
2820 XS = PH(2*I):GOSUB10040:P$=P$+XS$
2830 P$=STR$(I)+P$:GOSUB50000
2840 NEXT I
2850 REM  **   CALCULATE FOR EACH MEMBER:
2860 REM         MEMBER ELONGATION,
2865 REM         FORCES AND STRESSES.              **
2870 L1=1
2880 I=0
2890 GOSUB50010
2900 P$=" M       ELON           FORCE          STRESS"
2905 GOSUB 50000
2910 I=I+1
2920 J2=2*NJ(I)
2930 I2=2*NI(I)
2940 II=I2/2
2950 JJ=J2/2
2960 DU=PH(J2-1)-PH(I2-1)
2970 DV=PH(J2)-PH(I2)
2980 GOTO 3140
2990 REM  **   ELONGATION CALCULATED BY COMBINING
2995 REM         NETT COORDINATE DISPLACEMENTS OF
3000 REM         ONE END RELATIVE TO THE OTHER.     **
3010 EL=DU*CL+DV*CM
3020 REM  **   FORCE CALCULATED FROM STRESS/STRAIN
3025 REM         RELATION AND ELONGATION.           **
3030 REM THEN STRESS =   FORCE/AREA.
3040 FC=EL*E(I)*A(I)/DL
3050 RS =FC/A(I)
3060 FW=15:NS=4
3070 XS = EL:GOSUB10040:P$=XS$
3080 XS = FC:GOSUB10040:P$=P$+XS$
3090 XS = RS:GOSUB10040:P$=P$+XS$
3100 P$=STR$(I)+P$:GOSUB50000
3110 IF I-NM <0 GOTO 2910
3120 IF I-NM>=0 GOTO 3240
3130 REM  **   DETERMINE MEMBER LENGTHS DL AND
3135 REM         DIRECTION COSINES CL,CM..         **
3140 DX = X(JJ)-X(II)
3150 DY=Y(JJ)-Y(II)
3160 DL=SQR(DX↑2+DY↑2)
3170 CL=DX/DL
3180 CM=DY/DL
3190 IF L1 = 0 GOTO1180
3200 GOTO 3010
3210 REM  **   MULTIPLY UNCONDENSED STIFFNESS
3220 REM         MATRIX [KH] BY THE EXPANDED
3225 REM         (UNCONDENSED) DISPLACEMENT VECTOR [PH]
3230 REM         TO FIND ALL NODAL FORCES.          **
3240 FOR I=1 TO N
3250 FOR J=1 TO N
3260 F(I) =KH(I,J)*PH(J)+F(I)
```

```
3270 NEXT J
3280 NEXT I
3290 GOSUB50010
3300 P$="NODE          FX              FY":GOSUB50000
3310 FW=15:NS=4
3320 FOR I = 1 TO NN:P$=STR$(I)+"     "
3330 XS = F(2*I-1):GOSUB10040:P$=P$+XS$
3340 XS = F(2*I)  :GOSUB10040:P$=P$+XS$
3350 GOSUB50000
3360 NEXT I
3365 GOSUB 52000 :REM CLOSE PRINTER CHANNEL
3380 PRINT"DO YOU WISH TO RERUN THE PROGRAM USING"
3390 PRINT"  THE SAME # OF NODES AND MEMBERS BUT"
3400 PRINT"  -WITH DIFFERENT DATA FOR THE NODES"
3410 PRINT "     AND/OR MEMBERS? [Y/N]  ";N$;L4$;
3415 INPUT AN$
3420 IF LEFT$(AN$,1)<>"Y" THEN 3450
3430 IR = 1
3440 GOTO 350
3450 P$="        **** END OF RUN OF PROGRAM PJFRAME ****"
3455 GOSUB51000:IP=3:GOSUB50020
3460 GOSUB50000:IP=5:GOSUB50020
3470 END
10000 REM              FORMATTING AND INPUT/OUTPUT
10001 REM                  SUBROUTINES BY
10002 REM                   DAVID A. PIRIE
10003 REM     DEPARTMENT OF AERONAUTICS & FLUID MECHANICS
10004 REM                 UNIVERSITY OF GLASGOW
10005 REM                      SCOTLAND
10006 REM                   AUGUST  1983
10010 REM
10015 REM  **  FORMAT NUMERICAL OUTPUT IN
10020 REM       SCIENTIFIC NOTATION              **
10035 FW=12:NS=4
10040 WE =1E-30
10045 KE=0:KE$="":BL$="          ":B0$="00000000"
10050 F5=FW-NS-5:N3=NS+3:Z$="0.":AX=ABS(XS)
10052 IF AX<WE THEN XS$=LEFT$(BL$,F5)+Z$+LEFT$(BL$,N3):GOTO10095
10055 IFABS(XS)<.01ORABS(XS)>=1E9THEN10080
10060 IFABS(XS)<10ORABS(XS)>=10THENGOSUB10175
10065 GOSUB10110
10070 GOTO10095
10080 XS$=STR$(XS):KE$=RIGHT$(XS$,3):KE=VAL(KE$)
10085 XS=VAL(LEFT$(XS$,LEN(XS$)-4))
10090 GOSUB10110
10095 RETURN
10110 REM FORM O/P$
10115 GOSUB10145
10120 IFABS(XS)>=10THENGOSUB10175
10125 GOSUB10200
10130 GOSUB10225
10135 RETURN
10145 REM ROUNDOFF MANTISSA
10155 XR=5:FORI5=1TONS:XR=XR/10:NEXTI5
10160 XS=XS+XR*SGN(XS)
10165 RETURN
10175 REM NORMALISE MANTISSA
10180 IF ABS(XS)<1THENXS=XS*10:KE=KE-1:GOSUB10180
10185 IF ABS(XS)>=10THENXS=XS/10:KE=KE+1:GOSUB10185
10190 RETURN
```

```
10200 REM FORM EXPONENT$
10205 S$="+":IFKE<0THENS$="-"
10210 KE$=S$+RIGHT$("0"+MID$(STR$(KE),2),2)
10215 RETURN
10225 REM FORM (MANT+EXP)$
10230 X1$=LEFT$(STR$(XS),NS+2)
10235 XS$=X1$+LEFT$(B0$,NS+2-LEN(X1$))
10240 IFXS=INT(XS)THEN XS$=X1$+"."+LEFT$(B0$,NS-1)
10245 XS$=LEFT$(BL$,FW-NS-6)+XS$+"E"+KE$
10250 RETURN
20000 REM  **   FORMAT NUMERICAL OUTPUT            **
20020 FW=12:NS=3
20040 BL$="                "
20050 XS$=STR$(XS):XE$="    ":IFLEN(XS$)<4THENXS$=XS$+"    "
20060 IFABS(XS)>=10↑(8-NS)THENXX=XS:GOTO20080
20070 XX=XS+.5*SGN(XS)/10↑NS
20080 IFMID$(XS$,LEN(XS$)-3,1)="E"THENGOSUB20180
20090 XX$=STR$(XX)
20100 FORJ5=1TOLEN(XX$)
20110 IFMID$(XX$,J5,1)="."THENDP=J5:GOTO20130
20120 NEXTJ5:DP=J5:XX$=XX$+".0000000"
20130 XS$=LEFT$(XX$,DP+NS)+XE$
20140 LX=LEN(XS$):IFLX>FWTHENXS$=LEFT$(XS$,FW):GOTO20160
20150 XS$=LEFT$(BL$,FW-LX)+XS$
20160 RETURN
20180 XE$=RIGHT$(XS$,4):XR=VAL(RIGHT$(XS$,2))
20190 XX=VAL(LEFT$(XS$,LEN(XS$)-4))+.5*SGN(XS)/10↑NS
20200 RETURN
50000 REM ** THE FOLLOWING STATEMENTS
50001 REM    MUST BE TAILORED TO THE
50002 REM     PARTICULAR MACHINE IN USE          **
50005 PRINT#5,P$:RETURN :REM ** PRINTLINE ON PRINTER **
50010 PRINT#5:RETURN     :REM ** 1 LINEFEED ON PRINTER **
50015 REM
50020 FOR KP = 1 TO IP :REM IP
50021 PRINT#5          :REM    LINEFEEDS
50022 NEXT KP          :REM    ON
50023 RETURN           :REM    PRINTER *
50025 REM
51000 OPEN 5,4:RETURN  :REM ** OPEN CHANNEL TO PRINTER  **
52000 CLOSE 5:RETURN   :REM ** CLOSE CHANNEL TO PRINTER **
53000 REM ** THE FOLLOWING Y$,N$,L4$,AK$
53001 REM    ARE USED WITH 'INPUT' STATEMENTS -
53002 REM    SET THEM ALL EQUAL TO "" IF
53003 REM    L4$ NOT POSSIBLE ON YOUR MACHINE      **
53010 Y$="Y ":N$="N ":AK$="* "
53020 L4$="▮▮▮▮":RETURN :REM ** L4$ = 4 'CURSOR-LEFTS'  **
```

2.3.4 Example of truss analysis using *PJFRAME*

(a) *The structure:* A roof truss with assorted point loads.

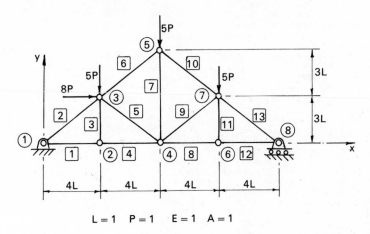

L = 1   P = 1   E = 1   A = 1

(b) *The data.* Data is fed in through the screen in an interactive mode.

(c) *The solution*.

```
****      PROGRAM PJFRAME      ****
```

N	KODE	X	Y	PX	PY
1	1	0.000	0.000	0.000	0.000
2	4	4.000	0.000	0.000	0.000
3	4	4.000	3.000	8.000	-5.000
4	4	8.000	0.000	0.000	0.000
5	4	8.000	6.000	0.000	-5.000
6	4	12.000	0.000	0.000	0.000
7	4	12.000	3.000	0.000	-5.000
8	2	16.000	0.000	0.000	0.000

M	NI	NJ	AREA	E
1	1	2	1.000	1.000E+00
2	1	3	1.000	1.000E+00
3	2	3	1.000	1.000E+00
4	2	4	1.000	1.000E+00
5	3	4	1.000	1.000E+00
6	3	5	1.000	1.000E+00
7	4	5	1.000	1.000E+00
8	4	6	1.000	1.000E+00
9	4	7	1.000	1.000E+00
10	5	7	1.000	1.000E+00
11	6	7	1.000	1.000E+00
12	6	8	1.000	1.000E+00
13	7	8	1.000	1.000E+00

TOP LEFT 6X6 CORNER OF THE CONDENSED STIFFNESS MATRIX

```
 5.000E-01   0.          0.          0.          -2.500E-01   0.

 0.          3.333E-01   0.         -3.333E-01    0.          0.

 0.          0.          3.840E-01   9.600E-02   -1.280E-01   9.600E-02

 0.         -3.333E-01   9.600E-02   5.493E-01    9.600E-02  -7.200E-02

-2.500E-01   0.         -1.280E-01   9.600E-02    7.560E-01   0.

 0.          0.          9.600E-02  -7.200E-02    0.          3.107E-01
```

```
          CONDENSED LOAD VECTOR
                 0.000
                 0.000
                 8.000
                -5.000
                 0.000
                 0.000
                 0.000
                -5.000
                 0.000
                 0.000
                 0.000
                -5.000
                 0.000
```

```
VECTOR OF UNKNOWN DISPLACEMENTS
      6.400E+01
     -3.611E+02
      2.083E+02
     -3.611E+02
      1.280E+02
     -3.918E+02
      1.276E+02
     -3.438E+02
      1.760E+02
     -3.397E+02
      6.294E+01
     -3.397E+02
      2.240E+02
```

```
VECTOR OF ALL NODAL DISPLACEMENTS
 N       UX              VY
 1     0.              0.
 2     6.400E+01       -3.611E+02
 3     2.083E+02       -3.611E+02
 4     1.280E+02       -3.918E+02
 5     1.276E+02       -3.438E+02
 6     1.760E+02       -3.397E+02
 7     6.294E+01       -3.397E+02
 8     2.240E+02        0.
```

```
 M       ELON            FORCE           STRESS
 1     6.400E+01        1.600E+01        1.600E+01
 2    -5.000E+01       -1.000E+01       -1.000E+01
 3     0.              0.               0.
 4     6.400E+01        1.600E+01        1.600E+01
 5    -4.583E+01       -9.167E+00       -9.167E+00
 6    -5.417E+01       -1.083E+01       -1.083E+01
 7     4.800E+01        8.000E+00        8.000E+00
 8     4.800E+01        1.200E+01        1.200E+01
 9    -2.083E+01       -4.167E+00       -4.167E+00
10    -5.417E+01       -1.083E+01       -1.083E+01
11     0.              0.               0.
12     4.800E+01        1.200E+01        1.200E+01
13    -7.500E+01       -1.500E+01       -1.500E+01
```

```
NODE     FX              FY
 1    -8.000E+00        6.000E+00
 2     0.              0.
 3     8.000E+00       -5.000E+00
 4     2.980E-08       -5.215E-08
 5    -1.490E-08       -5.000E+00
 6    -1.490E-08        0.
 7     0.              -5.000E+00
 8     0.               9.000E+00
```

**** END OF RUN OF PROGRAM PJFRAME ****

**Appendix 2.1   PJFRAME   (Computer Program for Plane Pin-jointed Frame
Analysis: program summary and data sheet**

1. *Introduction*

From the joint and member data read into the program, the stiffness matrix for
each member is evaluated and added into the complete (global) stiffness
matrix. The order of the coefficients in this global matrix is according to the
displacements u_1, v_1, u_2, $v_2 \ldots$ etc. The general member stiffness matrix is
formulated as follows:

$$X_i = \frac{EA}{L}\left[l^2 u_i + lm v_i - l^2 u_j - lm v_j\right]$$

$$Y_i = \frac{EA}{L}\left[lm u_i + m^2 v_i - lm u_j - m^2 v_j\right]$$

$$X_j = \frac{EA}{L}\left[-l^2 u_i - lm v_i + l^2 u_j + lm v_j\right]$$

$$Y_j = \frac{EA}{L}\left[-lm u_i - m^2 v_i + lm u_j + m^2 v_j\right]$$

or

For beam length L,
E = Young's Modulus
A = Cross-section area
$l = \cos\theta$, $m = \sin\theta$, direction cosines

$$\frac{EA}{L}\begin{bmatrix} l^2 & lm & -l^2 & -lm \\ lm & m^2 & -lm & -m^2 \\ -l^2 & -lm & l^2 & lm \\ -lm & -m^2 & lm & m^2 \end{bmatrix}\begin{bmatrix} u_i \\ v_i \\ u_j \\ v_j \end{bmatrix} = \begin{bmatrix} X_i \\ Y_i \\ X_j \\ Y_j \end{bmatrix}$$

Once the last member stiffness matrix has been added into the complete
stiffness matrix, this global stiffness matrix is reduced in size (condensed) to the
minimum number of equations, relating the unknown nodal displacements to
the corresponding nodal forces. Nodes can be restrained in any of the following
4 ways.

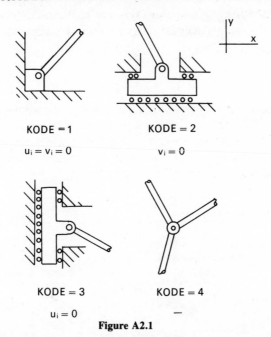

Figure A2.1

The solution is found as a set of nodal displacements, which are then used to calculate the elongation of, the force on and the stress in each member. The data required to be read in is detailed in section 2.3.2. An example of data is shown in section 2.3.4. The preparation of input data for this program should be accomplished in the following sequence.

(a) Sketch the structure and number the joints and members.

(b) Establish an origin of coordinates and label joints with proper coordinate values.

(c) Define the different load cases to be considered.

(d) Fill out datasheet.

2. Datasheet for **PJFRAME**

(*Note:* units must be consistent).

Data and prompts for **PJFRAME** (done in interrogative mode on the micro screen).

How many members in the structure?						
How many nodes?						

Nodal data is required

Node number?	1	2	3	4	5	6
Type in kode (see sheet)						
X & Y coords of node?	,	,	,	,	,	,
Applied X & Y forces at node?	,	,	,	,	,	,
Any more nodes or corrections?	YES/NO	YES/NO	YES/NO	YES/NO	YES/NO	YES/NO

Do you wish a print-out of nodal data?	YES/NO
Do you wish to correct the data?	YES/NO

Member data is required

Member number, and I & J end nodes	1, ,	2, ,	3, ,	4, ,	5, ,	6, ,
Member area and elastic modulus?	,	,	,	,	,	,
Any more members or corrections?	YES/NO	YES/NO	YES/NO	YES/NO	YES/NO	YES/NO

Do you wish a print-out of member data?	YES/NO
Do you wish to correct the data?	YES/NO

The program is now running

Do you wish the program to continue?	YES/NO	once during output
Do you wish to change data and rerun?	YES/NO	

Footnote: This is merely a sample blank data sheet for up to 6 nodes and 6 elements.

3 PLFRAME: Rigid/pin-jointed plane frames

3.0 Introduction

Analysis of pin-jointed frames or trusses can be thought of as an ideal or limiting condition in structural analysis, since no rotational restraints are supplied at the frictionless pin joints and consequently no moments are transmitted through the joint from member to member. This latter simplification results in only two unknowns or degrees of freedom at each node, viz. u and v. In order to model the other limiting condition, it is assumed that the nodal joints are rigid, have determinate rotations, and transmit bending moments from member to member. The inclusion of moments insinuates that the members are now beams and thus the corresponding stiffness method of analysis is called *beam element analysis*. There is now an extra degree of freedom at each joint, a rotation, with an extra equation relating moment equilibrium.

Whereas for the truss, the member stiffness matrix was a 4 × 4, the beam element member stiffness matrix becomes a 6 × 6. Because of this extra complexity, the derivation is done in two steps: (i) the stiffness matrix is found relative to the member's own or local coordinate system; (ii) it is then 'transformed' into global coordinates relative to the structure's coordinate system.

The development of the equations in section 3.1 is done in this order as indeed is the implementation in the program **PLFRAME**.

3.1 Development of equations

3.1.1 *Stiffness matrix about member axis*

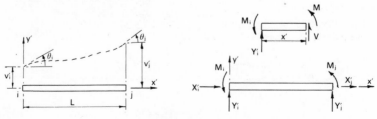

Figure 3.1

Consider the uniform beam ij of length L shown in Fig. 3.1. It carries no transverse loading along its length but *it is forced* to take up prescribed displacements v_i and v_j at its ends i and j and prescribed slopes or rotations θ_i and θ_j. The displacements are with reference to a *local* coordinate system x′ y′ with its origin at the i end of the beam. The rotations θ are about the z axis which is normal to the plane of the page.

From the moment curvature relation and the geometry of deformation, for linear elastic materials

$$d^2v'/dx'^2 = M/EI$$

where M is the moment at any section x′, E is the modulus of elasticity and I is the second moment of area of the beam cross section.

From Fig. 3.1, it can be seen that at section x′, $M = Y'_i x' - M_i$, by equilibrium.

Thus
$$\frac{d^2v'}{dx'^2} = \frac{1}{EI}(Y'_i x' - M_i)$$

Integration gives

$$\frac{dv'}{dx'} = \frac{1}{EI}(Y'_i \frac{x'^2}{2} - M_i x') + C_1 \tag{3.1}$$

and

$$v' = \frac{1}{EI}(Y'_i \frac{x'^3}{6} - \frac{M_i x'^2}{2}) + C_1 x' + C_2 \tag{3.2}$$

The boundary conditions are:

at
$$x' = 0, \frac{dv'}{dx'} = \theta_i, v' = v'_i \tag{3.3}$$

and at
$$x' = L, \frac{dv'}{dx'} = \theta_j, v' = v'_j \tag{3.4}$$

Thus substitution of (3.3) into equation (3.1) and (3.2) leads to

$$C_1 = \theta_i \text{ and } C_2 = v'_i$$

and substitution of (3.4) into (3.1) and (3.2) leads to

$$\theta_j = \frac{1}{EI}(Y'_i \frac{L^2}{2} - M_i L) + \theta_i \tag{3.5}$$

$$v'_j = \frac{1}{EI}(Y'_i \frac{L^3}{6} - M_i \frac{L^2}{2}) + \theta_i L + v'_i \tag{3.6}$$

Substitution of the expression for M_i from (3.5) into (3.6) gives

$$Y'_i = \frac{6EI}{L^2}(\theta_i + \theta_j) - \frac{12EI}{L^3}(v'_j - v'_i) \tag{3.7}$$

and back substitution leads to

$$M_i = \frac{EI}{L}(4\theta_i + 2\theta_j) - \frac{6EI}{L^2}(v'_j - v'_i) \tag{3.8}$$

Equilibrium applied to the beam gives two more equations $Y'_j = -Y'_i$

and $M_j = -Y'_j L - M_i$

and so substitution of (3.7) and (3.8) leads to

$$Y'_j = -\frac{6EI}{L^2}(\theta_i + \theta_j) + \frac{12EI}{L^3}(v'_j - v'_i) \tag{3.9}$$

$$M_j = \frac{EI}{L}(2\theta_i + 4\theta_j) - \frac{6EI}{L^2}(v'_j - v'_i) \tag{3.10}$$

For completeness, consideration of the horizontal forces X'_i and X'_j and their associated displacements u'_i and u'_j lead to two more equations

$$X'_i = -X'_j = \frac{EA}{L}(u'_i - u'_j) \tag{3.11}$$

where A is the area of the beam cross-section.

Equations (3.7) to (3.11) can be rewritten as follows

$$X'_i = \frac{EA}{L}u'_i + 0 + 0 - \frac{EA}{L}u'_j + 0 + 0$$

$$Y'_i = 0 + \frac{12EI}{L^3}v'_i + \frac{6EI}{L^2}\theta_i + 0 - \frac{12EI}{L^3}v_j + \frac{6EI}{L^2}\theta_j$$

$$M_i = 0 + \frac{6EI}{L^2} + \frac{4EI}{L}\theta_i + 0 - \frac{6EI}{L^2}v'_j + \frac{2EI}{L}\theta_j$$

$$X'_j = -\frac{EA}{L}u'_i + 0 + 0 + \frac{EA}{L}u_j' + 0 + 0$$

$$Y'_j = 0 - \frac{12EI}{L^3}v'_i - \frac{6EI}{L^2}\theta_i + 0 + \frac{12EI}{L^3}v'_j - \frac{6EI}{L^2}\theta_j$$

$$M_j = 0 + \frac{6EI}{L^2}v'_i + \frac{2EI}{L}\theta_i + 0 + \frac{6EI}{L^2}v'_j + \frac{4EI}{L}\theta_j$$

or more tidily put into matrix form –

$$
\begin{bmatrix} X'_i \\ Y'_i \\ M_i \\ X'_j \\ Y'_j \\ M_j \end{bmatrix}
=
\begin{bmatrix}
\dfrac{EA}{L} & 0 & 0 & -\dfrac{EA}{L} & 0 & 0 \\[2ex]
0 & \dfrac{12EI}{L^3} & \dfrac{6EI}{L^2} & 0 & -\dfrac{12EI}{L^3} & \dfrac{6EI}{L^2} \\[2ex]
0 & \dfrac{6EI}{L^2} & \dfrac{4EI}{L} & 0 & -\dfrac{6EI}{L^2} & \dfrac{2EI}{L} \\[2ex]
-\dfrac{EA}{L} & 0 & 0 & \dfrac{EA}{L} & 0 & 0 \\[2ex]
0 & -\dfrac{12EI}{L^3} & -\dfrac{6EI}{L^2} & 0 & \dfrac{12EI}{L^3} & -\dfrac{6EI}{L^2} \\[2ex]
0 & \dfrac{6EI}{L^2} & \dfrac{2EI}{L} & 0 & -\dfrac{6EI}{L^2} & \dfrac{4EI}{L}
\end{bmatrix}
\begin{bmatrix} u'_i \\ v'_i \\ \theta_i \\ u'_j \\ v'_j \\ \theta_j \end{bmatrix}
\qquad (3.12a)
$$

As with the pin-jointed structure, the above equations relate the applied member end forces and moments to the nodal displacements and rotations, thus giving an indicaton of the stiffness of the member. The 6×6 matrix of coefficients is called the stiffness matrix relative to the member's own axis system (in local coordinates) and called $[\mathbf{K'}]$.

$$\text{Thus } \{\mathbf{P'}\} = [\mathbf{K'}]\{\mathbf{U'}\} \qquad (3.13)$$

where $\{\mathbf{P'}\}$ is the load vector and $\{\mathbf{U'}\}$ the displacement vector relative to the local coordinate system.

Note that $[\mathbf{K'}]$ is square and symmetrical.

3.1.2 *Application of equations (3.12) to a simple orthogonal structure*

In the structure shown in Fig. 3.2 the members ABC and BD are of the same material and cross-section A. The members are rigidly joined together at B and the structure is encastré at C and D.

Obtain the equations of equilibrium for the nodes A and B under the loading system. From these equations and using the equations (3.12) obtain the structure stiffness matrix $[\mathbf{K'_1}]$ in the equation

$$\{W'\} = [K'_1] \{D'\}$$

where $\{D'\}$ is the vector of unknown displacements and $\{W'\}$ the load vector of the structure.

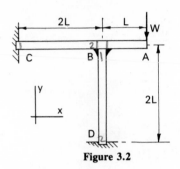

Figure 3.2

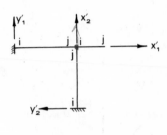

Figure 3.3

(i)   Boundary conditions:-

$$\left. \begin{array}{l} \text{At C: } u'_{1C} = v'_{1C} = \theta_C = 0 \\ \text{At D: } u'_{2D} = v'_{2D} = \theta_D = 0 \end{array} \right\} (\alpha)$$

– referring to the local coordinate systems $x'_1\ y'_1$ and $x'_2\ y'_2$ shown in Fig. 3.3.

Matching at B leads to

$$\left. \begin{array}{l} u'_{2B} = v'_{1B} \\ v'_{2B} = -u'_{1B} \end{array} \right\} (\beta)$$

The unknowns are thus $u'_{1A}, v'_{1A}, \theta_A, u'_{1B}, v'_{1B}, \theta_B$.

(ii)   Equilibrium (Fig. 3.4):

At A $(X'_A =)$ $X'_{AB} =$ $0 - (a)$

$(Y'_A =)$ $Y'_{AB} = -W - (b)$

$(M_A =)$ $M_{AB} =$ $0 - (c)$

At B $(X'_B =)$ $X'_{BA} + X'_{BC} - Y'_{BD} = 0 - (d)$

$(Y'_B =)$ $Y'_{BA} + Y'_{BC} + X'_{BD} = 0 - (e)$

$(M_B =)$ $M_{BA} + M_{BC} + M_{BD} = 0 - (f)$

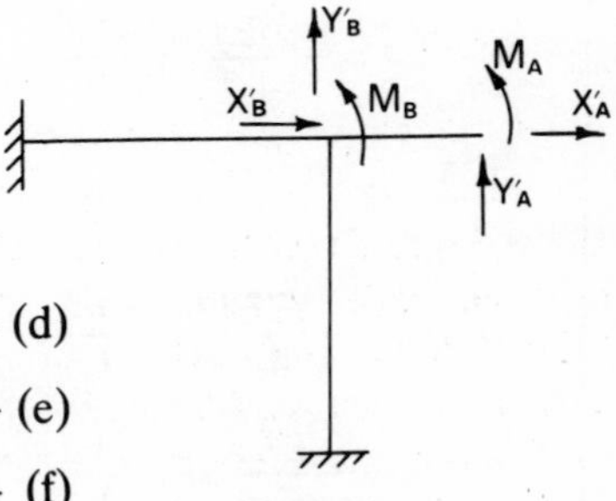

Figure 3.4

C

(iii)   The equations (3.12) are now applied to the equilibrium equations (a) –
        (f) using the boundary conditions (α). All displacements are referred to
        $x'_1 \, y'_1$ system of axes by use of conditions (β).

(a) $X'_A = \dfrac{EA}{L} u'_{1A} - \dfrac{EA}{L} u'_{1B}$

(b) $Y'_A = \dfrac{12EI}{L^3} v'_{1A} - \dfrac{6EI}{L^2} \theta_A - \dfrac{12EI}{L^3} v'_{1B} - \dfrac{6EI}{L^2} \theta_B$

(c) $M_A = -\dfrac{6EI}{L^2} v'_{1A} + \dfrac{4EI}{L} \theta_A + \dfrac{6EI}{L^2} v'_{1B} + \dfrac{2EI}{L} \theta_B$

(d) $X'_B = -\dfrac{EA}{L} u'_{1A} + (\dfrac{EA}{L} + \dfrac{EA}{2L} + \dfrac{12EI}{(2L)^3}) u'_{1B} + \dfrac{6EI}{(2L)^2} \theta_B$

(e) $Y'_B = -\dfrac{12EI}{L^3} v'_{1A} + \dfrac{6EI}{L^2} \theta_A + (\dfrac{12EI}{L^3} + \dfrac{12EI}{(2L)^3} + \dfrac{EA}{2L}) v'_{1B}$

$$+ (\dfrac{6EI}{L^2} - \dfrac{6EI}{(2L)^2}) \; \theta_B$$

(f) $M_B = -\dfrac{6EI}{L^2} v'_{1A} + \dfrac{2EI}{L} \theta_A + \dfrac{6EI}{(2L)^2} u'_{1B} + (\dfrac{6EI}{L^2} - \dfrac{6EI}{(2L)^2}) v'_{1B}$

$$+ (\dfrac{4EI}{L} + \dfrac{4EI}{2L} + \dfrac{4EI}{2L}) \theta_B$$

The stiffness matrix $[K'_1]$ is formed by collecting together the coefficients of
the displacement components, such that

u'_{1A}	v'_{1A}	θ_A	u'_{1B}	v'_{1B}	θ_B
$\dfrac{EA}{L}$	0	0	$-\dfrac{EA}{L}$	0	0
0	$\dfrac{12EI}{L^3}$	$-\dfrac{6EI}{L^2}$	0	$-\dfrac{12EI}{L^3}$	$-\dfrac{6EI}{L^2}$
0	$-\dfrac{6EI}{L^2}$	$\dfrac{4EI}{L}$	0	$\dfrac{6EI}{L^2}$	$\dfrac{2EI}{L}$
$-\dfrac{EA}{L}$	0	0	$\dfrac{3}{2}(\dfrac{EA}{L} + \dfrac{EI}{L^3})$	0	$\dfrac{3}{2}\dfrac{EI}{L^2}$
0	$-\dfrac{12EI}{L^3}$	$\dfrac{6EI}{L^2}$	0	$\dfrac{27}{2}\dfrac{EI}{L^3} + \dfrac{EA}{2L}$	$\dfrac{9}{2}\dfrac{EI}{L^2}$
0	$-\dfrac{6EI}{L^2}$	$\dfrac{2EI}{L}$	$\dfrac{3}{2}\dfrac{EI}{L^2}$	$\dfrac{9}{2}\dfrac{EI}{L^2}$	$8\dfrac{EI}{L}$

Note the symmetry of the structure matrix.

The load vector $\{W'\}$ is given from

$$
\begin{bmatrix}
X'_A \\
Y'_A \\
M_A \\
X'_B \\
Y'_B \\
M_B
\end{bmatrix}
=
\begin{bmatrix}
0 \\
-W \\
0 \\
0 \\
0 \\
0
\end{bmatrix}
\qquad \text{from equations (a)} \rightarrow \text{(f).}
$$

Given specific values of L, A, I, E and W, the solution to the six simultaneous equations can be found longhand, by using equation solution procedures such as Gaussian elimination or by inverting matrix $[K'_1]$. There are many computer programs to solve simultaneous linear algebraic equations and these can easily be employed to give the solution.

$$\text{Thus } \{D'\} = [K'_1]^{-1}\{W'\}$$

The structure stiffness matrix $[K'_1]$ was 'built up' without requiring knowledge of the loading, which is incorporated into the matrix $\{W'\}$. Complex loading will alter $\{W'\}$ but leave the stiffness matrix unchanged.

e.g. the loading indicated in Fig. 3.5 would give

$$
\{W'\} =
\begin{bmatrix}
-H_1 \\
-W_1 \\
-Q_1 \\
0 \\
+W_2 \\
+Q_2
\end{bmatrix}
$$

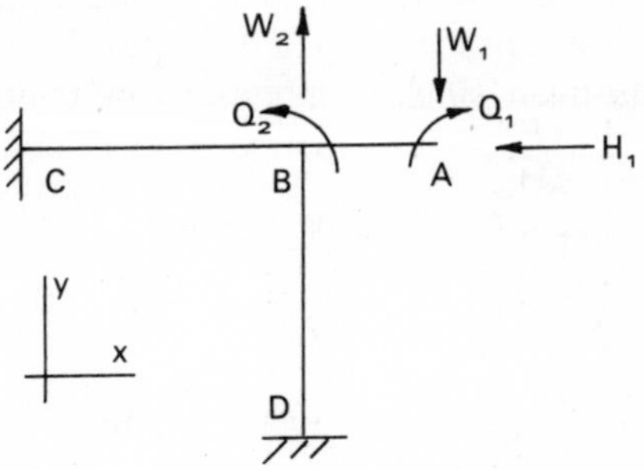

Figure 3.5

3.1.3 *Transforming the member stiffness matrix $[K']$ from local to global coordinates*

The statement of the member stiffness with reference to its own coordinate system is too restrictive. For example in section 3.1.2, if member BD had been inclined, the matching condition (β) at B would be more troublesome. It is easier to refer all nodal displacements and forces to a common structure or global axis system, thus requiring a transformation of the stiffness matrix as follows.

Fig. 3.6 shows the j end of member ij which is originally in a position indicated by $i_0 \, j_0$ but deforms to $i_1 \, j_1$. Small displacements are assumed and so the angle θ of $i_0 \, j_0$ to the global axis x does not alter significantly. The local axis system is given by x', y' with its origin at i_0. The node j in moving from j_0 to j_{1-} has coordinate displacements (u'_j, v'_j) or (u_j, v_j) with respect to either local or global axes.

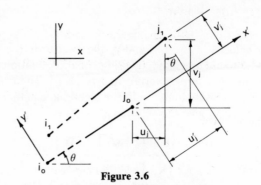

Figure 3.6

Thus
$$u_j = u'_j \cos\theta - v'_j \sin\theta$$
$$\text{and } v_j = + u'_j \sin\theta + v'_j \cos\theta \tag{3.14}$$

Note that the rotation θ_j is about the z axis normal to the plane of the paper and is consequently unaffected by the orientation of the x-y axes. At the i end of the beam similar relations can be found such that the transformation matrix is

$$
\begin{bmatrix} u_i \\ v_i \\ \theta_i \\ u_j \\ v_j \\ \theta_j \end{bmatrix}
=
\begin{bmatrix}
\cos\theta & -\sin\theta & 0 & 0 & 0 & 0 \\
\sin\theta & \cos\theta & 0 & 0 & 0 & 0 \\
0 & 0 & 1 & 0 & 0 & 0 \\
0 & 0 & 0 & \cos\theta & -\sin\theta & 0 \\
0 & 0 & 0 & \sin\theta & \cos\theta & 0 \\
0 & 0 & 0 & 0 & 0 & 1
\end{bmatrix}
\begin{bmatrix} u'_i \\ v'_i \\ \theta_i \\ u'_j \\ v'_j \\ \theta_j \end{bmatrix}
$$

$$\tag{3.15}$$

$$\text{or } \{U\} = [T]\,\{U'\} \tag{3.16}$$

where $[T]$ is the transformation matrix
and $\{U\}$, $\{U'\}$ are the displacement vectors
with respect to the x-y and x'-y' axes respectively.

Multiplying both sides of (3.16) by the inverse* of $[T]$ gives

$$[T]^{-1} \{U\} = \{U'\} \tag{3.17}$$

However, referring back to Fig. 3.6, the relations between u', v' and u, v could also be written as

$$u'_j = u_j \cos\theta + v_j \sin\theta$$

$$v'_j = -u_j \sin\theta + v_j \cos\theta$$

and similarly at node i. Thus the equations equivalent to (3.15) are

$$
\begin{bmatrix} u'_i \\ v_i \\ \theta_i \\ u'_j \\ v'_j \\ \theta_j \end{bmatrix}
=
\begin{bmatrix}
\cos\theta & \sin\theta & 0 & 0 & 0 & 0 \\
-\sin\theta & \cos\theta & 0 & 0 & 0 & 0 \\
0 & 0 & 1 & 0 & 0 & 0 \\
0 & 0 & 0 & \cos\theta & \sin\theta & 0 \\
0 & 0 & 0 & -\sin\theta & \cos\theta & 0 \\
0 & 0 & 0 & 0 & 0 & 1
\end{bmatrix}
\begin{bmatrix} u_i \\ v_i \\ \theta_i \\ u_j \\ v_j \\ \theta_j \end{bmatrix}
\tag{3.18}
$$

$$or \ \{U'\} = [T^i] \{U\} \tag{3.19}$$

Comparison of (3.19) and (3.17) gives

$$[T^i] = [T]^{-1}$$

but comparison of (3.18) with (3.15) shows that

$$[T]^{-1} = [T]^T \tag{3.20}$$

---

* The product $[T]^{-1} [T] = [I]$ the unit matrix

$$
\begin{bmatrix}
1 & 0 & 0 & 0 & 0 & 0 \\
0 & 1 & 0 & 0 & 0 & 0 \\
0 & 0 & 1 & 0 & 0 & 0 \\
0 & 0 & 0 & 1 & 0 & 0 \\
0 & 0 & 0 & 0 & 1 & 0 \\
0 & 0 & 0 & 0 & 0 & 1
\end{bmatrix}
$$

which is equivalent to multiplying by unity.

In an analogous way the applied end forces to the beam can be transformed. Reference to Fig. 3.7 will show that at end j

$$X_j = X'_j \cos\theta - Y'_j \sin\theta$$

$$Y_j = X_j \sin\theta + Y_j \cos\theta$$

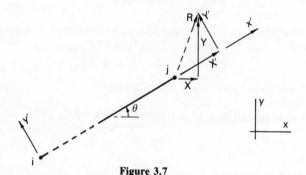

Figure 3.7

The resultant end force R_j remains unchanged such that

$$X'^2 + Y'^2 = X^2 + Y^2 = R^2$$

As with θ_i, θ_j, M_i and M_j are with respect to the z axis.

And so
$$\{P\} = [T]\{P'\} \qquad (3.21)$$

or
$$[T]^{-1}\{P\} = \{P'\} \qquad (3.22)$$

where $[T]$ is the same transformation matrix as before,

$$\{P\} = \begin{bmatrix} X_i \\ Y_i \\ M_i \\ X_j \\ Y_j \\ M_j \end{bmatrix} \quad \text{and} \quad \{P'\} = \begin{bmatrix} X'_i \\ Y'_i \\ M_i \\ X'_j \\ Y'_j \\ M_j \end{bmatrix}$$

are the load vectors with respect to the x – y and x' – y' axes respectively.

The transformation matrix has mainly terms in $\cos\theta$ and $\sin\theta$, which are the direction cosines l and m, and so $[T]$ could be written

$$[T] = \left[\begin{array}{ccc|ccc} l & -m & 0 & 0 & 0 & 0 \\ m & l & 0 & 0 & 0 & 0 \\ 0 & 0 & 1 & 0 & 0 & 0 \\ \hline 0 & 0 & 0 & l & -m & 0 \\ 0 & 0 & 0 & m & l & 0 \\ 0 & 0 & 0 & 0 & 0 & 1 \end{array}\right]$$

(3.23)

Returning to equation

$$\{P'\} = [K']\{U'\}$$ (3.13)

and using $[T]^{-1}\{U\} = \{U'\}$ (3.17)

and $[T]^{-1}\{P\} = \{P'\}$ (3.22)

leads to $[T]^{-1}\{P\} = [K'][T]^{-1}\{U\}$.

Pre-multiplying both sides by $[T]$ leads to

$$\{P\} = [T][K'][T]^{-1}\{U\}$$ (3.24)

As $[T]^{-1} = [T]^{T}$, equation (3.20), and since the transpose is more easily found than the inverse, (3.24) is usually written as

$$\{P\} = [T][K'][T]^{T}\{U\}$$ (3.25)

$$or \{P\} = [K]\{U\}$$ (3.26)

where $[K] = [T][K'][T]^{T}$ is the member global stiffness matrix i.e. written with respect to the global or structure coordinate system.

In computer programs $[K]$ is not usually explicitly stored, but calculated for each member from $[K']$ and $[T]$.

However for the example in the next section the matrix $[K]$ is explicitly given as (see Fig. 3.8.)

$$[K] = \frac{EA}{L}
\begin{array}{cccccc}
u_i & v_i & \theta_i & u_j & v_j & \theta_j \\
\end{array}$$

$$[K] = \frac{EA}{L}
\begin{bmatrix}
l^2 + m^2\,\dfrac{12k^2}{L^2} & lm - lm\,\dfrac{12k^2}{L^2} & -m\,\dfrac{6k^2}{L} & -l^2 - m^2\,\dfrac{12k^2}{L^2} & -lm + lm\,\dfrac{12k^2}{L^2} & -m\,\dfrac{6k^2}{L} \\[2ex]
lm - lm\,\dfrac{12k^2}{L^2} & m^2 + l^2\,\dfrac{12k^2}{L^2} & l\,\dfrac{6k^2}{L} & -lm + lm\,\dfrac{12k^2}{L^2} & -m^2 - l^2\,\dfrac{12k^2}{L^2} & l\,\dfrac{6k^2}{L} \\[2ex]
-m\,\dfrac{6k^2}{L} & l\,\dfrac{6k^2}{L} & 4k^2 & m\,\dfrac{6k^2}{L} & -l\,\dfrac{6k^2}{L} & 2k^2 \\[2ex]
-l^2 - m^2\,\dfrac{12k^2}{L^2} & -lm + lm\,\dfrac{12k^2}{L^2} & m\,\dfrac{6k^2}{L} & l^2 + m^2\,\dfrac{12k^2}{L^2} & -lm - lm\,\dfrac{12k^2}{L^2} & m\,\dfrac{6k^2}{L} \\[2ex]
-lm + lm\,\dfrac{12k^2}{L^2} & -m^2 - l^2\,\dfrac{12k^2}{L^2} & -l\,\dfrac{6k^2}{L} & lm - lm\,\dfrac{12k^2}{L^2} & m^2 + l^2\,\dfrac{12k^2}{L^2} & -l\,\dfrac{6k^2}{L} \\[2ex]
-m\,\dfrac{6k^2}{L} & l\,\dfrac{6k^2}{L} & 2k^2 & m\,\dfrac{6k^2}{L} & -l\,\dfrac{6k^2}{L} & 4k^2
\end{bmatrix}$$

(3.27)

I = uniform second moment of cross-section

a = uniform cross-section area

$k^2 = \sqrt{I}/a$ = radius of gyration of cross section

(l, m): direction of cosines of member axis Ox' relative to 'frame axes' Ox, Oy.

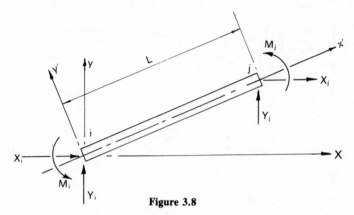

Figure 3.8

3.1.4 Example

Determine the stiffness matrix of the given structure.

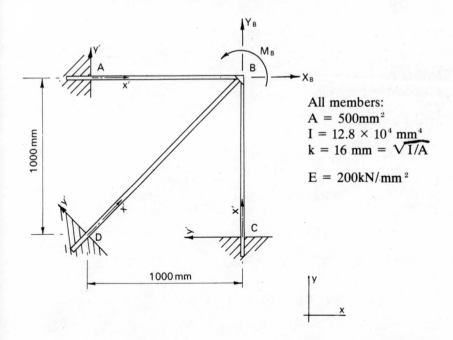

All members:
A = 500mm^2
I = 12.8 × 10^4 mm^4
k = 16 mm = $\sqrt{I/A}$

E = 200kN/mm^2

Boundary conditions: $u_A = v_A = \theta_A = u_C = u_C = \theta_C = u_D = v_D = \theta_D = 0$

Non-zero displacement components: u_B, u_B, θ_B

Member	L	l	m	l^2	m^2	lm	$\dfrac{EA}{L}$ kN/mm
AB	1000	1	0	1	0	0	100
CB	1000	0	1	0	1	0	100
DB	$1000\sqrt{2}$	$\dfrac{1}{\sqrt{2}}$	$\dfrac{1}{\sqrt{2}}$	$\dfrac{1}{2}$	$\dfrac{1}{2}$	$\dfrac{1}{2}$	$50\sqrt{2}$

Note:
$$\frac{EA}{L_2} = \frac{1}{\sqrt{2}} \frac{EA}{L_1}$$

The stiffness matrix for the structure is

$$[\mathbf{K}] = \begin{bmatrix} k_{11} & & \text{Sym} \\ k_{21} & k_{22} & \\ k_{31} & k_{32} & k_{33} \end{bmatrix}$$

Either of two methods may be used.

Method (1) To form the elements of the matrix put one component of displacement = 1 and others equal to zero.

$u_B = 1, u_B = \theta_B = 0$

For k_{11}, $X_B = X_{BA} + X_{BC} + X_{BD}$

$$= 100 \times 1 + 100 \times \frac{12 \times 16^2}{1000^2} + 50\sqrt{2}\,(\tfrac{1}{2} + \tfrac{1}{2}\frac{12 \times 16^2}{2 \times 10^6})$$

$$= 135.717$$

For k_{21}, $Y_B = Y_{BA} + Y_{BC} + Y_{BD}$

$$= 0 + 0 + 50\sqrt{2}\,(\tfrac{1}{2} - \tfrac{1}{2}\ \frac{12 \times 16^2}{2 \times 10^6}) = 35.3007$$

For k_{31}, $M_B = M_{BA} + M_{BC} + M_{BD}$

$$= 0 + 100 \frac{6 \times 16^2}{10^3} + 50\sqrt{2} \times \frac{1}{\sqrt{2}} \frac{6 \times 16^2}{\sqrt{2} \times 10^3}$$

$$= 207.905$$

$v_B = 1$, $u_B = \theta_B = 0$

For k_{22}, $Y_B = Y_{BA} + Y_{BC} + Y_{BD}$

$$= 100 \times \frac{12 \times 16^2}{10^6} + 100 + 50\sqrt{2} \left(\tfrac{1}{2} + \tfrac{1}{2}\frac{12 \times 16^2}{2 \times 10^6}\right)$$

$$= 135.717$$

For k_{32}, $M_B = M_{BA} + M_{BC} + M_{BD}$

$$= 100 \times \left(-\frac{6 \times 16^2}{10^3}\right) + 0 + 50\sqrt{2} \left(-\frac{1}{\sqrt{2}} \frac{6 \times 16^2}{\sqrt{2} \times 10^3}\right)$$

$$= -207.905$$

$\theta_B = 1$, $u_B = v_B = 0$

For k_{33}, $M_B = M_{BA} + M_{BC} + M_{BD}$

$$= 100 \times 4 \times 16^2 + 100 \times 4 \times 16^2 + 50\sqrt{2} \times 4 \times 16^2$$

$$= 277,207.$$

Thus,

	u_B	v_B	θ_B
$[\mathbf{K}] =$	135.717	35.3007	207.905
	35.3007	135.717	−207.905
	207.905	−207.905	277 207

Note: Units employed are kN and mm; M_B has units kNmm.

Addendum

Inverting $[\mathbf{K}]$ gives flexibility matrix

$$
[\mathbf{K}]^{-1} = 10^{-3}
\begin{array}{c}
\begin{array}{ccc}
\quad X_B & \qquad Y_B & \qquad M_B
\end{array} \\
\begin{bmatrix}
+7.918 & -2.071 & -0.00749 \\
-2.071 & +7.918 & +0.00749 \\
-0.00749 & +0.00749 & +0.00362
\end{bmatrix}
\end{array}
$$

Method (2). Establish the stiffness matrix for each member and then combine by using equilibrium.

Only the relevant coefficients are given.

Member AB: $l = 1, m = 0$

$$
\begin{bmatrix}
X_{AB} \\
Y_{AB} \\
M_{AB} \\
X_{BA} \\
Y_{BA} \\
M_{BA}
\end{bmatrix}
=
\frac{EA}{L_1}
\begin{bmatrix}
- & - & - & - & - & - \\
- & - & - & - & - & - \\
- & - & - & - & - & - \\
- & - & - & 1 & 0 & 0 \\
- & - & - & 0 & \dfrac{12k^2}{L_1^2} & \dfrac{-6k^2}{L_1} \\
- & - & - & 0 & \dfrac{-6k^2}{L_1} & 4k^2
\end{bmatrix}
\begin{bmatrix}
u_A \\
v_A \\
\theta_A \\
u_B \\
v_B \\
\theta_B
\end{bmatrix}
$$

Member BD: $l = m = \sqrt{\dfrac{1}{2}}$

$$
\begin{bmatrix}
X_{DB} \\
Y_{DB} \\
M_{DB} \\
X_{BD} \\
Y_{BD} \\
M_{BD}
\end{bmatrix}
=
\frac{EA}{L_2}
\begin{bmatrix}
- & - & - & - & - & - \\
- & - & - & - & - & - \\
- & - & - & - & - & - \\
- & - & - & \tfrac{1}{2}+\tfrac{1}{2}\dfrac{12k^2}{L_2^2} & \tfrac{1}{2}-\tfrac{1}{2}\dfrac{12k^2}{L_2^2} & \sqrt{\tfrac{1}{2}}\dfrac{6k^2}{L_1} \\
- & - & - & \tfrac{1}{2}-\tfrac{1}{2}\dfrac{12k^2}{L_2^2} & \tfrac{1}{2}+\tfrac{1}{2}\dfrac{12k^2}{L_2^2} & -\sqrt{\tfrac{1}{2}}\dfrac{6k^2}{L_2} \\
- & - & - & \sqrt{\tfrac{1}{2}}\dfrac{6k^2}{L_2} & -\sqrt{\tfrac{1}{2}}\dfrac{6k^2}{L_2} & 4k^2
\end{bmatrix}
\begin{bmatrix}
u_D \\
v_D \\
\theta_D \\
u_B \\
v_B \\
\theta_B
\end{bmatrix}
$$

Member CB: $\quad l = 0, m = 1$

$$
\begin{bmatrix} X_{CB} \\ Y_{CB} \\ M_{CB} \\ X_{BC} \\ Y_{BC} \\ M_{BC} \end{bmatrix} = \frac{EA}{L_1} \begin{bmatrix} - & - & - & - & - & - \\ - & - & - & - & - & - \\ - & - & - & - & - & - \\ - & - & - & \dfrac{12k^2}{L_1{}^2} & 0 & \dfrac{6k^2}{L_1} \\ - & - & - & 0 & 1 & 0 \\ - & - & - & \dfrac{6k^2}{L_1} & 0 & 4k^2 \end{bmatrix} \begin{bmatrix} u_C \\ v_C \\ \theta_C \\ u_B \\ v_B \\ \theta_B \end{bmatrix}
$$

Equilibrium gives

$$X_B = X_{BA} + X_{BC} + X_{BD}$$
$$Y_B = Y_{BA} + Y_{BC} + Y_{BD}$$
$$M_B = M_{BA} + M_{BC} + M_{BD}$$

The matrix on p.70 follows,

where

$$k_{11} = k_{22} = \frac{EA}{L_1} \left(1 + \frac{1}{\sqrt{2}} \left(\tfrac{1}{2} \frac{6.256}{10^6 .2} \right) + \frac{12.256}{10^6} \right) = 135.717$$

$$k_{21} = \frac{EA}{L_1} \left(\tfrac{1}{2} - \frac{6.256}{10^6 .2} \right) \frac{1}{\sqrt{2}} = 35.3007$$

$$k_{31} = -k_{32} = \frac{EA}{L_1} \left(\frac{3.256}{1000. \sqrt{2}} + \frac{6.256}{1000} \right) = 207.905$$

$$k_{33} = \frac{EA}{L_1} \left(4.256 + \frac{1}{\sqrt{2}} \cdot 4.256 + 4.256 \right) = 277207.0$$

$$\begin{Bmatrix} X_B \\ Y_B \\ M_B \end{Bmatrix} = \frac{EA}{L_1} \begin{bmatrix} u_B & v_B & \theta_B \end{bmatrix}$$

Column 1 (u_B):

$$(1)+\left(\tfrac{1}{2}+\tfrac{1}{2}\cdot\frac{12k^2}{L_2^2}\right)\frac{1}{\sqrt{2}}+\left(\frac{12k^2}{L_1^2}\right) = k_{11}$$

$$(0)+\left(\tfrac{1}{2}-\tfrac{1}{2}\cdot\frac{12k^2}{L_2^2}\right)\frac{1}{\sqrt{2}}+(0) = k_{21}$$

$$(0)+\left(\frac{1}{\sqrt{2}}\cdot\frac{6k^2}{L_2}\right)\frac{1}{\sqrt{2}}+\frac{6k^2}{L_1} = k_{31}$$

Column 2 (v_B):

$$k_{12} = k_{21}$$

$$\left(\frac{12k^2}{L_1^2}\right)+\left(\tfrac{1}{2}+\tfrac{1}{2}\cdot\frac{12k^2}{L_2^2}\right)\frac{1}{\sqrt{2}}+(1) = k_{22}$$

$$\left(-\frac{6k^2}{L_2}\right)+\left(-\frac{1}{\sqrt{2}}\cdot\frac{6k^2}{L_2}\right)\frac{1}{\sqrt{2}}+(0) = k_{32}$$

Column 3 (θ_B):

$$k_{13} = k_{31}$$

$$k_{23} = k_{32}$$

$$(4k^2)+(4k^2)\frac{1}{\sqrt{2}}+(4k^2) = k_{33}$$

ie: $\{P\} = [K]\,\{U\}$

3.1.5 *Dealing with different end conditions*

In order to deal with structures in which all the joints are not rigid but in which some are pinned, different local stiffness matrices are developed in the same way as was shown in section 3.1.1. In all cases the matrix relative to the local coordinate system, $[K']$, is developed. In order to find the global stiffness matrix $[K]$, the triple product in equation (3.25) is found.

When a pin-joint exists there is zero moment transmitted from member to member and the rotation θ has no meaning since the members, which are attached to that pin-joint, each rotate by different amounts.

Thus in the following matrices the rows and columns corresponding to such pin joints are omitted.

Type 0 = fixed/fixed

$$[K'] = \begin{array}{c}\begin{array}{cccccc} u'_i & v'_i & \theta_i & u'_j & v'_j & \theta_j \end{array} \\ \left[\begin{array}{cccccc} \dfrac{EA}{L} & 0 & 0 & -\dfrac{EA}{L} & 0 & 0 \\[2mm] 0 & \dfrac{12EI}{L^3} & \dfrac{6EI}{L^2} & 0 & -\dfrac{12EI}{L^3} & \dfrac{6EI}{L^2} \\[2mm] 0 & \dfrac{6EI}{L^2} & \dfrac{4EI}{L} & 0 & -\dfrac{6EI}{L^2} & \dfrac{2EI}{L} \\[2mm] -\dfrac{EA}{L} & 0 & 0 & \dfrac{EA}{L} & 0 & 0 \\[2mm] 0 & -\dfrac{12EI}{L^3} & -\dfrac{6EI}{L^2} & 0 & \dfrac{12EI}{L^3} & -\dfrac{6EI}{L^2} \\[2mm] 0 & \dfrac{6EI}{L^2} & \dfrac{2EI}{L} & 0 & -\dfrac{6EI}{L^2} & \dfrac{4EI}{L} \end{array} \right] \end{array} \quad (3.28)$$

– as found in equation (3.12*a*)

Type 1 – pinned/fixed

$$
[\mathbf{K'}] =
\begin{array}{c}
\quad u'_i \quad\quad v'_i \quad\quad \theta_i \quad\quad u'_j \quad\quad v'_j \quad\quad \theta_j \\
\left[
\begin{array}{cccccc}
\dfrac{EA}{L} & 0 & - & -\dfrac{EA}{L} & 0 & 0 \\[2.2ex]
0 & \dfrac{3EI}{L^3} & - & 0 & -\dfrac{3EI}{L^3} & \dfrac{3EI}{L^2} \\[2.2ex]
- & - & - & - & - & - \\[2.2ex]
-\dfrac{EA}{L} & 0 & - & \dfrac{EA}{L} & 0 & 0 \\[2.2ex]
0 & -\dfrac{3EI}{L^3} & - & 0 & \dfrac{3EI}{L^3} & -\dfrac{3EI}{L^2} \\[2.2ex]
0 & \dfrac{3EI}{L^2} & - & 0 & -\dfrac{3EI}{L^2} & \dfrac{3EI}{L}
\end{array}
\right]
\end{array}
\qquad (3.29)
$$

Type 2 – fixed/pinned

$$
[\mathbf{K'}] =
\begin{array}{c}
\quad u'_i \quad\quad v'_i \quad\quad \theta_i \quad\quad u'_j \quad\quad v'_j \quad\quad \theta_j \\
\left[
\begin{array}{cccccc}
\dfrac{EA}{L} & - & - & -\dfrac{EA}{L} & - & - \\[2.2ex]
 & \dfrac{3EI}{L^3} & \dfrac{3EI}{L^2} & 0 & -\dfrac{3EI}{L^3} & - \\[2.2ex]
0 & \dfrac{3EI}{L^2} & \dfrac{3EI}{L} & 0 & -\dfrac{3EI}{L^2} & - \\[2.2ex]
-\dfrac{EA}{L} & 0 & 0 & \dfrac{EA}{L} & 0 & - \\[2.2ex]
0 & -\dfrac{3EI}{L^3} & -\dfrac{3EI}{L^2} & 0 & \dfrac{3EI}{L^3} & - \\[2.2ex]
- & - & - & - & - & -
\end{array}
\right]
\end{array}
\qquad (3.30)
$$

Type 3 – pinned/pinned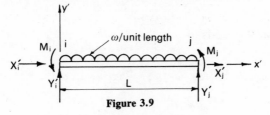

$$
[\mathbf{K}'] = \begin{array}{cccccc}
u'_i & v'_i & \theta_i & u'_j & v'_j & \theta_j \\
\dfrac{EA}{L} & - & - & -\dfrac{EA}{L} & - & - \\[2ex]
- & - & - & - & - & - \\[2ex]
- & - & - & - & - & - \\[2ex]
-\dfrac{EA}{L} & - & - & \dfrac{EA}{L} & - & - \\[2ex]
- & - & - & - & - & - \\[2ex]
- & - & - & - & - & -
\end{array}
$$

(3.31)

In local coordinates a pinned/pinned member can only transmit an axial (X') force and consequently sustain an axial (u') displacement.

3.1.6 *Dealing with member loads*

Two specific situations are considered – a beam with a uniformly distributed load and a beam with a point load. In each case, the equivalent nodal loads can be determined and so the lateral load on the beam 'shed' to the nodes and treated as applied nodal loads.

Uniformly distributed lateral load: As in section 3.1.1 consider the uniform beam ij of length L but now with a uniformly distributed transverse load ω/ unit length (Fig. 3.9); for other nomenclature refer to Fig. 3.1. Once again refer force and displacement action to the beam's own or local axis system $x'y'$.

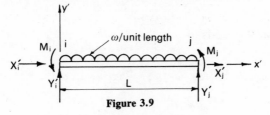

Figure 3.9

Using moment curvature as before gives

$$\frac{d^2v'}{} = \frac{1}{EI} \ (Y'_i \ x' - M_i - \frac{\omega x'^2}{2})$$

Integration gives $\frac{dv'}{dx'} = \frac{1}{EI} \ (Y'_i \ \frac{x'^2}{2} - M_i x' - \frac{\omega x'^3}{6}) + C_1$ \hfill (3.32)

and $v' = \frac{1}{EI} \ (Y'_i \ \frac{x'^3}{6} - M_i \ \frac{x'^2}{2} - \frac{\omega x'^4}{24}) + C_1 x + C_2$ \hfill (3.33)

As before the boundary conditions are:

$$\text{at } x' = 0, \frac{dv'}{dx'} = \theta_i, \ v' = v'_i \hfill (3.34)$$

$$\text{and at } x' = L, \frac{dv'}{dx'} = \theta_j, \ v' = v'_j \hfill (3.35)$$

Substitution of (3.34) into (3.32) and (3.33) leads to

$$C_1 = \theta_i \text{ and } C_2 = v'_i$$

and substitution of (3.35) into (3.32) and (3.33) leads to

$$\theta_j = \frac{1}{EI} \ (Y'_i \frac{L^2}{2} - M_i L - \frac{\omega L^3}{6}) + \theta_i \hfill (3.36)$$

$$v'_j = \frac{1}{EI} \ (Y'_i \frac{L^3}{6} - M_i \frac{L^2}{2} - \frac{\omega L^4}{24}) + \theta_i L + v'_i \hfill (3.37)$$

Substitution of the expression for M_i from (3.36) into (3.37) gives

$$Y'_i = \frac{6EI}{L^2} \ (\theta_i + \theta_j) - \frac{12EI}{L^3} \ (v'_j - v'_i) + \frac{\omega L}{2} \hfill (3.38)$$

and back substitution leads to

$$M_i = \frac{EI}{L} \ (4\theta_i + 2\theta_j) - \frac{6EI}{L^2} \ (v'_j - v'_i) + \frac{\omega L^2}{12} \hfill (3.39)$$

Equilibrium gives the equations $Y'_j = -Y'_i + \omega L$

and $M_j = Y'_i L - M_i - \dfrac{\omega L^2}{2}$, thus leading to

$$Y'_j = -\frac{6EI}{L^2}(\theta_i + \theta_j) + \frac{12EI}{L^3}(v'_j - v'_i) + \frac{\omega L}{2} \qquad (3.40)$$

and $M'_j = \dfrac{EI}{L}(2\theta_i + 4\theta_j) - \dfrac{6EI}{L^3}(v'_j - v'_i) - \dfrac{\omega L^2}{12}$ \qquad (3.41)

The lateral loading leaves the expressions for the horizontal loads unaffected as before viz.

$$X'_i = -X'_j = \frac{EA}{L}(u'_i - u_j) \qquad (3.42)$$

Equations (3.38) to (3.42) are similar to the equations (3.12), the only difference being the terms in w which can be thought of as equivalent external loadings. To illustrate this consider the two-span beam ABC shown in Fig. 3.10, in which each span carries a different, but uniformly distributed load.

The beam is rigidly fixed at A and C.

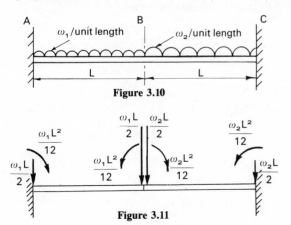

Figure 3.10

Figure 3.11

As in the example in section 3.1.2, equilibrium at node B gives

$$\left. \begin{array}{l} (X'_B =)\ X'_{BA} + X'_{BC} = 0 \\[4pt] (Y'_B =)\ Y'_{BA} + Y'_{BC} = 0 \\[4pt] (M_B =)\ M_{BA} + M_{BC} = 0 \end{array} \right\} \qquad (3.43)$$

Since the spans carry uniformly distributed loads ω_1 and ω_2, the special beam element equations (3.38) to (3.42) must be used in (3.43) to produce the three equations in the three unknowns u'_B, v'_B and θ_B. The terms in ω_1 and ω_2 would be transferred to the right-hand side of equations (3.43) and give

$$0 \ , \ (\frac{-\omega_1 L}{2} - \frac{\omega_2 L}{2}) \text{ and } (\frac{\omega_1 L^2}{12} - \frac{\omega_2 L^2}{12}) \text{ respectively.}$$ But this is equivalent

to using the original beam element equation (3.12) in (3.43) and applying equivalent nodal loads such that

$$X'_B = 0 \ , \ Y'_B = -\frac{\omega_1 L}{2} - \frac{\omega_2 L}{2} \ , \ M_B = \frac{\omega_1 L^2}{12} - \frac{\omega_2 L^2}{12}$$

Thus the situation illustrated in Fig. 3.11 will give the same solution as that in Fig. 3.10. The procedure can be thought of as *shedding* member loads to the nodes, and treating them as equivalent external nodal loads which are added to any other structure loads which are applied at the same node.

Should the member axes be inclined to the structure or global axes, the member's equivalent external nodal shear loads ($\omega L/2$) must be resolved into two components parallel to the global axes.

To summarize: if a structure has a member which carried a uniformly distributed lateral load ω, the equivalent external loads $\omega L/2$ and $\omega L^2/12$ should be calculated and applied to the structure as additional nodal loads (Fig. 3.12). Care should be taken in the directions of moments and forces, which should be resolved into the structure axes.

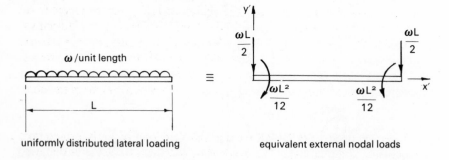

uniformly distributed lateral loading   equivalent external nodal loads

Figure 3.12

Point load: Two approaches can be adopted. First, the point of application of the load on a beam can be treated as a node and a rigid connection assumed between the sections of the beam of either side. The member load then simply becomes a nodal load. This approach has the one disadvantage of creating another node, which increases the storage requirement of the program.

Alternatively, an analysis similar to that above can be performed and the point load shed to the nodes as equivalent external nodal loads. This is left as an exercise for the reader.

3.2 Implementation in program

3.2.1 *Description of program*

The program **PLFRAME** is given in full in section 3.2.3 below. The data must be inserted into the program in DATA statements between lines 4000 and 10000 in the order indicated in section 3.2.2. This form of data input means that the data is held within the program and can be stored on disk or tape for future running and correction if required.

Stiffness matrices. The program scans through each member and calculates its local stiffness matrix from (3.28), (3.29), (3.30) or (3.31) depending on whether the member is of type MT = 0,1, 2 or 3. The geometry of the member, then permits the forming of the transformation matrix $[\mathbf{T}]$, (sometimes called the rotation matrix). In order to establish the member global stiffness matrix $[\mathbf{KG}]$ the triple product $[\mathbf{T}] [\mathbf{KL}] [\mathbf{T}]^{\mathrm{T}}$ is established, and held in $[\mathbf{KH}]$ for future use. In contrast to the program **PJFRAME,** the matrix is located directly into the condensed structure stiffness matrix. This is done by sensing the 'flag' RE which has three values at each node, corresponding to the three degrees of freedom (u, v, θ) at that node. If at node I, the flag RE (I, 1) = 1, this indicates that $u_I = 0$ and need not be considered in the condensed equations. Consequently any reference to u_I in the equations $[\mathbf{KG}]$ for the member are ignored and not transferred across to $[\mathbf{KS}]$. Similarly RE (I, 2) = 1 indicates v_I = 0 and RE (I, 3) = 1 that $\theta_I = 0$. The locations within this condensed matrix $[\mathbf{KS}]$ are complex and use is made of the index MN to give the row corresponding to a particular degree of freedom. The value of MN is calculated near the beginning of the program for each free (non-restrained) degree of freedom.

Load vector. As with **PJFRAME** the uncondensed load vector is formed from the applied nodal load data. The condensing process is the same as that used before and the condensed load vector is held in $\{\mathbf{P}\}$.

Solution procedure. With $[\mathbf{KS}]$ and $\{\mathbf{P}\}$ there are now the required number of unknown displacements – the unknown degrees of freedom. These linear algebraic simultaneous equations are solved by Gaussian elimination in the identical procedure to that in **PJFRAME** and the vector of unknown nodal displacements is held in the vector $\{\mathbf{P}\}$.

Member end loads and moments with respect to global coordinates. The program scans through each member and recalls its global stiffness matrix, which was stored as $\left[\textbf{KH}\right]$. From the known nodal displacements and rotations, the member's *own* displacement vector $\{\textbf{U}\}$ can be established. From (3.26) it can be seen that the product

$$\left[\textbf{KH}\right]\{\textbf{U}\} = \{\textbf{P}\}$$

will give the member load vector $\{\textbf{P}\}$, which contains the forces X, Y and moment M at each end i and j of the beam. Members of types 1, 2, 3 will of course leave M = 0 at the end(s) which are pinned, as no moment can be transmitted through this assumed friction-free joint.

Nodal forces. As with **PJFRAME** the applied nodal forces and reactions can be found by multiplying the uncondensed structure stiffness matrix by the uncondensed displacement vector. In **PLFRAME**, all members are scanned and all stiffness coefficients are added from member matrix $\left[\textbf{KH}\right]$ into structure matrix $\left[\textbf{KS}\right]$ (now uncondensed). The product of $\left[\textbf{KS}\right]$ with the vector of *all* nodal displacements will give for each node two forces and one moment, which are either applied forces and moments at unrestrained nodes (many of which will be zero, at unloaded nodes) or reaction forces and moments at restrained nodes, where the corresponding displacements or rotations are zero.

Note that this particular section of the program could be omitted and the reaction forces and moments calculated by applying equilibrium at boundary nodes of all members attached thereto. However, the whole procedure does act as a useful check on accuracy and conditioning of the equations. The sum of all X- and Y- nodal forces should be zero. Moment equilibrium of all nodal forces and moments will also be satisfied.

2.2.2   Data preparation

A data preparation sheet is found in Appendix 3.1. The data is required in the following order and where more than one item is required for entry, a comma is used as a separator.

Structure data.
(i)   Number of nodes in the structure, NN.
       This includes all reaction points.
(ii)   Number of members in the structure, NM.
Note that as written the program will solve for plane frames of up to 10 members (MX), 10 nodes (NX) and 30 degrees of freedom. To change these limits, MX and NX should be changed.

Nodal data. For each node,
(iii)   Node number – start first node as ① and there should be no gap to NN.

(iv) Coordinates x, y of the node.

(v) Restraints, RE (I, 1) = 1 for u_I = 0, else = 0

RE (I, 2) = 1 for v_I = 0, else = 0

RE (I, 3) = 1 for θ_1 = 0, else = 0.

(Note: if node I is a pin joint RE (I, 3) is always zero).

(vi) Applied nodal loads

X – direction force

Y – direction force

M – moment

Member data. For each member,

(vii) Member number – start first member as $\boxed{1}$ and there should be no gaps to NM.

(viii) The i and j end nodes.

(ix) Member type = 0 for fixed/fixed

= 1 for i end pinned/j end fixed

= 2 for i end fixed/j end pinned

= 3 for both ends pinned.

(x) Member properties: I – second moment of area of cross-section

A – area of cross-section

E – modulus of elasticity.

Notes (a) The member numbers and nodal numbers should start off at one and end with the last number with no gaps.

(b) Units are as decided by the user but must be consistent. For example if E is in units of N/mm^2, then the coordinates x and y are in mm, area A in mm^2, second moment of area in mm^4, forces in newtons and moments in Nmm. The output is also in the corresponding units, rotations being in radians.

(c) It is recommended that member and nodal data be checked carefully when printed out, before allowing the program to proceed.

*3.2.3 Program **PLFRAME***

Flow chart for **PLFRAME**

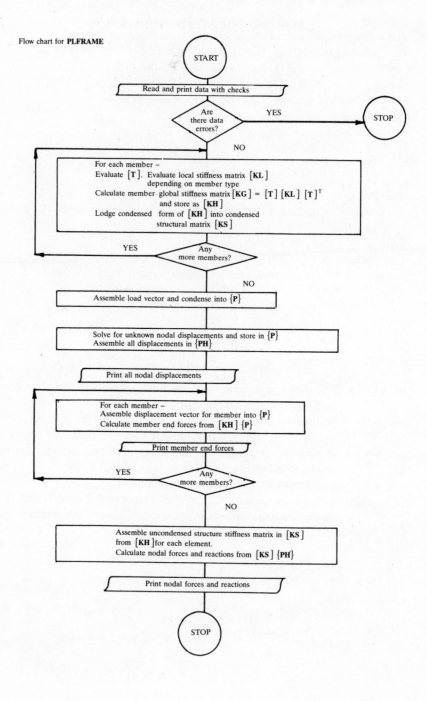

START

Read and print data with checks

Are there data errors? — YES → STOP

NO

For each member –
Evaluate $[\mathbf{T}]$. Evaluate local stiffness matrix $[\mathbf{KL}]$
depending on member type
Calculate member global stiffness matrix $[\mathbf{KG}] = [\mathbf{T}][\mathbf{KL}][\mathbf{T}]^{\mathrm{T}}$
and store as $[\mathbf{KH}]$
Lodge condensed form of $[\mathbf{KH}]$ into condensed
structural matrix $[\mathbf{KS}]$

YES ← Any more members?

NO

Assemble load vector and condense into $\{\mathbf{P}\}$

Solve for unknown nodal displacements and store in $\{\mathbf{P}\}$
Assemble all displacements in $\{\mathbf{PH}\}$

Print all nodal displacements

For each member –
Assemble displacement vector for member into $\{\mathbf{P}\}$
Calculate member end forces from $[\mathbf{KH}]\{\mathbf{P}\}$

Print member end forces

YES ← Any more members?

NO

Assemble uncondensed structure stiffness matrix in $[\mathbf{KS}]$
from $[\mathbf{KH}]$ for each element.
Calculate nodal forces and reactions from $[\mathbf{KS}]\{\mathbf{PH}\}$

Print nodal forces and reactions

STOP

```
100 REM              *** PROGRAM  PLFRAME ***
105 REM                AUTHOR:
110 REM                   DAVID K. BROWN
112 REM         DEPARTMENT OF MECHANICAL ENGINEERING
115 REM                UNIVERSITY OF GLASGOW
120 REM                     SCOTLAND
125 REM
126 REM                  AUGUST  1983
127 REM
130 REM   **   THE PROGRAM WILL ANALYSE PLANE FRAME
131 REM        PROBLEMS WITH UP TO 10 NODES,
132 REM        10 MEMBERS AND 30 DEGREES OF FREEDOM.
133 REM        LOADING IS APPLIED THROUGH
134 REM        NODAL FORCES AND MOMENTS.            **
135 REM
140 REM   **   FOUR TYPES OF MEMBERS ARE PERMITTED
150 REM        DEPENDING ON END CONDITIONS:-
160 REM            TYPE 0 - FIXED/FIXED
170 REM            TYPE 1 - PINNED/FIXED
180 REM            TYPE 2 - FIXED/PINNED
190 REM            TYPE 3 - PINNED/PINNED
200 REM   **   BOUNDARY CONDITIONS CAN BEFIXED (U=V=ANGLE=0)
201 REM        OR PINNED (U=V=0)OR A COMBINATION
202 REM        OF PINS AND ROLLERS.                **
205 REM
210 REM   **   INPUT TO THE PROGRAM IS FROM
211 REM        DATA STATEMENTS BETWEEN
212 REM        LINES 4000 AND 10000.              **
215 REM
220 REM   **   OUTPUT CONSISTS OF:
221 REM          NODAL DISPLACEMENTS U,V AND ROTATION,
222 REM          FORCES (X,Y) AND MOMENT AT EACH END
223 REM          OF EACH MEMBER AND NODAL FORCES
224 REM          (INCLUDING REACTIONS) AT EACH NODE **
225 REM
230 PL$=" PROGRAM PLFRAME "
235 GOSUB 53000 :REM MACHINE SPECIFIC STRINGS
240 NX=10:REM MAX. NO OF NODES = 10
250 MX=10:REM MAX. NO OF MEMBERS = 10
260 NF=3*NX:REM MAX # OF DEGREES OF FREEDOM
270 DIM X(NX),Y(NX),RE(NX,3),NJ(MX,2),MT(MX),MI(MX)
280 DIM A(MX),E(MX),QL(MX),T(6,6),U(6,6),KL(6,6)
290 DIM KG(6,6),KS(NF,NF),P(NF+2),PH(NF),ER(6)
300 DIM KH(MX,6,6),MN(NX,3)
310 BL$="                "
320 GOSUB51000:REM OPEN PRINTER CHANNEL
335 FORI=1TONF:FORJ=1TONF:KS(I,J)=0:NEXTJ:NEXTI
340 REM DATA READ IN FROM DATA STATEMENTS
350 READ NN,NM
360 FORI=1TO NN
370 READ N,X(I),Y(I),RE(I,1),RE(I,2),RE(I,3)
380 READ P(3*I-2),P(3*I-1),P(3*I):NEXTI
390 FOR I=1TO NM
400 READ N,NJ(I,1),NJ(I,2),MT(I),MI(I),A(I),E(I):NEXTI
410 P$=PL$:GOSUB50000:IP=3:GOSUB50020
420 P$="            *** DATA INPUT ***"
425 GOSUB50000:IP=2:GOSUB50020
430 P$="NUMBER OF NODES   = "+STR$(NN):GOSUB50000
```

```
440 P$="NUMBER OF MEMBERS = "+STR$(NM):GOSUB50000
450 IP=2:GOSUB50020:P$="* NODAL DATA *"
455 GOSUB50000:GOSUB50010
460 FW = 12:NS = 3
465 REM   **  VALUES FOR FORMATTING SUBROUTINE -
466 REM        FW MUST BE >=NS+7                      **
470 S1$=LEFT$(BL$,FW-1):S2$=LEFT$(BL$,3*FW-25)
475 S3$=LEFT$(BL$,FW-5)
480 P$="NODE       COORDINATES"+S2$
481 P$=P$+"R   E   S   T   R   A   I   N   T   S"
485 GOSUB 50000
490 P$=" NO       X"+S1$+"Y"+S1$+"U"
491 P$=P$+S1$+"V"+S3$+"A N G L E"
495 GOSUB 50000
500 FOR I=1TONN:P$=STR$(I)+"    "
510 XS=X(I):GOSUB20040: P$=P$+XS$
520 XS=Y(I):GOSUB20040: P$=P$+XS$
530 P$=P$+"   "; FOR J = 1 TO 3
540 P$ = P$ + STR$(RE(I,J))+"          "
550 NEXT J
560 GOSUB50000:NEXT I
570 IP=2:GOSUB50020:P$="* MEMBER DATA *"
575 GOSUB50000:GOSUB50010
580 FW=12:NS=3:REM VALUES FOR OUTPUT FORMATTING
590 P$="MEMBER END NODES     MEMBER      2ND"
591 P$=P$+" MOMT      SECTN        ELASTIC"
595 GOSUB50000
600 P$="  NO    I       J       TYPE       OF"
601 P$=P$+" AREA      AREA       MODULUS"
605 GOSUB50000
610 B$="       ":FORI=1TONM
615 P$="  "+STR$(I)+"      "+STR$(NJ(I,1))+"     "
616 P$=P$+STR$(NJ(I,2))+"      "+STR$(MT(I))+"     "
620 XS=MI(I):FW=11:NS=3:GOSUB10040:P$=P$+XS$
630 XS=A(I):GOSUB10040:P$=P$+XS$
640 XS=E(I):GOSUB10040:P$=P$+XS$
650 GOSUB50000:NEXTI
660 GOSUB50010:P$="* APPLIED NODAL FORCES *"
665 GOSUB50000:GOSUB50010
670 FW=12:NS=3
680 P$="NODE        F  O  R  C  E  S        MOMENT"
685 GOSUB 50000
690 P$=" NO       X              Y             M"
695 GOSUB 50000
700 FORI=1TONN:P$ = STR$(I) + "     "
710 XS=P(3*I-2):GOSUB20000:P$=P$+XS$
720 XS=P(3*I-1):GOSUB20000:P$=P$+XS$
730 XS=P(3*I):GOSUB20000:P$=P$+XS$
740 GOSUB50000:NEXTI
745 IP=5:GOSUB50020
750 GOSUB 52000: REM CLOSE PRINTER CHANNEL
755 PRINT "DO YOU WISH TO CORRECT THE DATA [Y/N]   ";N$;L4$;
756 INPUT AN$
760 IF LEFT$(AN$,1)="Y" THEN 2590
770 IF LEFT$(AN$,1)<>"N" THEN 660
780 PRINT "THE PROGRAM IS NOW RUNNING"
790 GOSUB 51000
800 IN = 0: FOR I=1TONN :FORJ=1TO3
810 IF RE(I,J)=1 THEN GOTO 830
820 IN= IN+1: MN(I,J)=IN
```

```
830 NEXT J
840 NEXTI
850 MK=IN
860 IF IN>1 THENGOTO890
870 GOSUB 52000
875 PRINT "PROGRAM WILL NOT RUN WITH ONLY"
876 PRINT "   ONE UNKNOWN DISPLACEMENT"
880 STOP
890 REM  **   CONDENSED STRUCTURAL STIFFNESS MATRIX
895 REM        IS MK X MK IN SIZE.                  **
900 REM  **   NOW SCAN THROUGH ALL MEMBERS          **
910 FOR IJK= 1 TO NM
920 FOR I=1TO6: FOR J=1TO6
930 T(I,J)=0:KL(I,J)=0:KG(I,J)=0:U(I,J)=0
940 NEXT J
950 NEXTI
960 REM  **   DETERMINE TRANSFORMATION MATRIX [T] **
970 II= NJ(IJK,1)
980 JJ= NJ(IJK,2)
990 XI=X(II): XJ=X(JJ)
1000 YI=Y(II): YJ=Y(JJ)
1010 QL(IJK)=SQR((XJ-XI)↑2+(YJ-YI)↑2)
1020 CX=(XJ-XI)/QL(IJK):CY=(YJ-YI)/QL(IJK)
1030 T(1,1)=CX:T(1,2)=CY:T(2,1)=-CY:T(2,2)=CX
1040 T(3,3)=1:T(4,4)=CX:T(4,5)=CY:T(5,4)=-CY
1050 T(5,5)=CX:T(6,6)=1
1060 REM  **   DETERMINE MEMBER LOCAL STIFFNESS
1065 REM        MATRIX [KL] DEPENDING ON
1070 REM        MEMBER TYPE  MT = 0,1,2 OR 3.       **
1080 C1=2*E(IJK)*MI(IJK)/QL(IJK)
1090 C2=3*C1/QL(IJK)
1100 C3=2*C2/QL(IJK)
1110 C4=E(IJK)*A(IJK)/QL(IJK)
1120 C5=1.5*C1
1130 C6=0.5*C2
1140 C7=C6/QL(IJK)
1150 KL(1,1)=C4:KL(1,4)=-C4:KL(4,1)=-C4
1160 KL(4,4)=C4
1170 IF MT(IJK)=3 THEN GOTO 1330
1180 IF MT(IJK)>0 THEN GOTO 1240
1190 KL(2,2)=C3:KL(3,2)=C2:KL(5,2)=-C3:KL(6,2)=C2
1200 KL(2,3)=C2:KL(3,3)=2*C1:KL(5,3)=-C2:KL(6,3)=C1
1210 KL(2,5)=-C3:KL(3,5)=-C2:KL(5,5)=C3:KL(6,5)=-C2
1220 KL(2,6)=C2:KL(3,6)=C1:KL(5,6)=-C2:KL(6,6)=2*C1
1230 GOTO 1330
1240 KL(2,2)=C7:KL(5,2)=-C7
1250 KL(2,5)=-C7:KL(5,5)=C7
1260 IF MT(IJK)=2 THEN GOTO 1300
1270 KL(2,6)=C6:KL(6,2)=C6:KL(6,6)=C5
1280 KL(5,6)=-C6:KL(6,5)=-C6
1290 GOTO 1330
1300 KL(2,3)=C6:KL(3,3)=C5:KL(5,3)=-C6
1310 KL(3,2)=C6:KL(3,5)=-C6
1320 REM  **   CALCULATE MEMBER GLOBAL STIFFNESS
1325 REM        MATRIX [KG] FROM TRIPLE PRODUCT.    **
1330 FORI=1TO6: FORJ=1TO6:FORL=1TO6
1340 U(I,J) = U(I,J)+KL(I,L)*T(L,J)
1350 NEXT L:NEXT J:NEXT I
1380 FORI=1TO6: FORJ=1TO6: FORL=1TO6
1390 KG(I,J) = KG(I,J)+T(L,I)*U(L,J)
```

$\{u\} = [T]\{u'\}$

(x is cos θ)

```
1400 NEXT L:NEXT J:NEXT I
1430 REM   **   STORE MEMBER STIFFNESS
1435 REM        MATRIX [KG] IN [KH].              **
1440 FORI=1TO6: FORJ=1TO6
1450 KH(IJK,I,J) = KG(I,J)
1460 NEXTJ:NEXTI
1470 REM   **   LOCATE STIFFNESS MATRIX COEFFICIENTS
1475 REM        IN STRUCTURAL STIFFNESS MATRIX [KS] **
1480 FORI=1TO2: ND=NJ(IJK,I): IS=3*I-2
1490 FORL=1TO3: IF RE(ND,L)=1 THEN GOTO1570
1500 PK=MN(ND,L)
1510 FORJ=1TO2: JS=3*J-2: NC=NJ(IJK,J)
1520 FORM=1TO3: IF RE(NC,M)=1 THEN GOTO1550
1530 PL=MN(NC,M)
1540 KS(PK,PL)=KS(PK,PL)+KG(IS+L-1,JS+M-1)
1550 NEXTM
1560 NEXTJ
1570 NEXTL
1580 NEXTI
1590 NEXTIJK
1600 REM   **   BUILD UP UNCONDENSED LOAD VECTOR
1605 REM        FROM DATA, THEN CONDENSE INTO [P]   **
1610 OT=0: FOR I=1TONN: FORJ=1TO3
1620 IF RE(I,J)=0 THEN GOTO 1660
1630 FORM=3*I-(3-J)-OT TO 3*NN-OT+1
1640 P(M)=P(M+1)
1650 NEXTM: OT=OT+1
1660 NEXTJ
1670 NEXTI
1680 REM   **   SOLVE FOR UNKNOWN DISPLACEMENTS
1685 REM        AND STORE IN [P].              **
1690 M=3*NN-OT: M1=M-1
1700 FORI=1TOM1 :L=I+1
1710 FORJ=LTOM
1720 IF KS(J,I)=0 THEN GOTO 1770
1730 FORKK=LTOM
1740 KS(J,KK)=KS(J,KK)-KS(I,KK)*KS(J,I)/KS(I,I)
1750 NEXTKK
1760 P(J)=P(J)-P(I)*KS(J,I)/KS(I,I)
1770 NEXTJ
1780 NEXTI
1790 P(M)=P(M)/KS(M,M)
1800 FORI=1TOM1: KK=M-I: L=KK+1
1810 FORJ=L TO M
1820 P(KK)=P(KK)-P(J)*KS(KK,J)
1830 NEXTJ
1840 P(KK)=P(KK)/KS(KK,KK)
1850 NEXTI
1860 IP=3:GOSUB50010
1865 P$="                  *** OUTPUT OF RESULTS ***"
1866 GOSUB 50000
1870 FORI=1TO30: PH(I)=0
1880 NEXTI
1890 IN=0: FORI=1TONN: FORJ=1TO3
1900 IF RE(I,J)=1 THEN GOTO 1920
1910 IN=IN+1: PH(3*I-3+J)=P(IN)
1920 NEXTJ
1930 NEXTI
1940 IP=3:GOSUB50010
1950 FW=15:NS=3
```

```
1960 P$="* VECTOR OF ALL DISPLACEMENTS *"
1965 GOSUB50000:GOSUB50010
1970 P$="NODE           DISPLACEMENTS                    ROTATION"
1975 GOSUB 50000
1980 P$=" NO            U                V              ANGLE"
1985 GOSUB50000:GOSUB50010
1990 FOR I=1 TO NN:P$=STR$(I)+"      "
1995 XS=PH(3*I-2):GOSUB20040:P$=P$+XS$
2000 XS=PH(3*I-1):GOSUB20040:P$=P$+XS$
2010 XS=PH(3*I):GOSUB20040:P$=P$+XS$
2020 GOSUB50000
2030 NEXTI
2040 REM  **   SCAN THROUGH ALL MEMBERS AND
2045 REM       CALCULATE APPLIED FORCES AND MOMENTS
2050 REM       AT THE END OF EACH MEMBER.          **
2052 REM
2055 REM  **   NOTE: DISPLACEMENTS ARE NOW
2060 REM       HELD IN VECTOR [P].                 **
2070 IP=2:GOSUB50010
2075 P$="* MEMBER END FORCES W.R.T GLOBAL COORDS *"
2076 GOSUB 50000:GOSUB 50010
2080 P$="MEMBER NODE          F O R C E S              MOMENT"
2085 GOSUB 50000
2090 P$=" NO    NO        X              Y              M"
2095 GOSUB 50000:GOSUB 50010
2100 FORIJK=1TONM
2110 FORI=1TO2: NB=NJ(IJK,I)
2120 FOR J=1TO3
2130 P(3*I-3+J)=PH(3*NB-3+J)
2140 NEXTJ:NEXTI
2150 FORI=1TO6:ER(I)=0:NEXTI
2160 FORI=1TO6:FORJ=1TO6
2170 ER(I)=ER(I)+KH(IJK,I,J)*P(J)
2180 NEXTJ:NEXTI
2185 FW=15:NS=3
2190 FORJ=1TO2:P$=" "+STR$(IJK)+"    "+STR$(NJ(IJK,J))+"   "
2200 XS=ER(3*J-2):GOSUB20040:P$=P$+XS$
2210 XS=ER(3*J-1):GOSUB20040:P$=P$+XS$
2220 XS=ER(3*J):GOSUB20040:P$=P$+XS$
2230 GOSUB50000:NEXTJ
2240 GOSUB50010
2250 NEXTIJK
2260 REM  **  BUILD UP STRUCTURE UNCONDENSED
2265 REM      STIFFNESS MATRIX AND HOLD IN [KS].  **
2270 FORI=1TO30: FORJ=1TO30
2280 KS(I,J)=0
2290 NEXTJ
2300 NEXTI
2310 FORIJK=1TONM
2320 FORI=1TO2: I1=3*NJ(IJK,I)-2: IS=3*I-2
2330 FORL=0TO2: FORJ=1TO2
2340 J1=3*NJ(IJK,J)-2: JS=3*J-2
2350 FORM=0TO2
2360 KS(I1+L,J1+M)=KS(I1+L,J1+M)+KH(IJK,IS+L,JS+M)
2370 NEXTM
2380 NEXTJ
2390 NEXTL
2400 NEXTI
2410 NEXTIJK
2420 FORI=1TO30:P(I)=0
```

```
2430 NEXTI
2440 REM   **   MULTIPLY [KS] BY VECTOR OF ALL NODAL
2445 REM        DISPLACEMENTS [PH] TO DETERMINE
2450 REM        NODAL FORCES AND THUS CHECK AGAINST
2455 REM        APPLIED LOADS AND FIND BOUNDARY
2460 REM        REACTION FORCES.                      **
2470 FORI=1TO3*NN:FORJ=1TO3*NN
2480 P(I)=P(I)+KS(I,J)*PH(J)
2490 NEXTJ
2500 NEXTI
2510 IP=2:GOSUB50020
2520 P$="* NODAL LOADS W.R.T. GLOBAL COORDS *"
2525 GOSUB 50000:GOSUB 50010
2530 P$="NODE          F O R C E S              MOMENT"
2535 GOSUB50000
2540 P$=" NO        X              Y            M"
2541 GOSUB50000
2545 FW=15:NS=3
2550 FOR J = 1 TO NN:P$ = STR$(J)+"  "
2560 FOR I = 1 TO 3
2570 XS = P(3*J-3+I):GOSUB20040:P$=P$+XS$
2580 NEXT I:GOSUB50000:IP=3:GOSUB50010:NEXT J
2585 P$="            **** END OF RUN OF PROGRAM PLFRAME ****"
2586 GOSUB 50000:IP=5:GOSUB50020
2590 GOSUB 52000
2600 END
2700 REM ****************************
4000 REM DATA STATEMENTS ARE LOCATED BETWEEN
4010 REM LINES 4000 AND 10000
4020 REM ***********************
10000 REM              FORMATTING AND INPUT/OUTPUT
10001 REM                    SUBROUTINES BY
10002 REM                   DAVID A. PIRIE
10003 REM    DEPARTMENT OF AERONAUTICS & FLUID MECHANICS
10004 REM                UNIVERSITY OF GLASGOW
10005 REM                     SCOTLAND
10006 REM                   AUGUST  1983
10010 REM
10015 REM   **  FORMAT NUMERICAL OUTPUT IN
10020 REM       SCIENTIFIC NOTATION                   **
10035 FW=12:NS=4
10040 WE =1E-30
10045 KE=0:KE$="":BL$="          ":B0$="00000000"
10050 F5=FW-NS-5:N3=NS+3:Z$="0. ":AX=ABS(XS)
10052 IF AX<WE THEN XS$=LEFT$(BL$,F5)+Z$+LEFT$(BL$,N3):GOTO10095
10055 IFABS(XS)<.01ORABS(XS)>=1E9THEN10080
10060 IFABS(XS)<1ORABS(XS)>=10THENGOSUB10175
10065 GOSUB10110
10070 GOTO10095
10080 XS$=STR$(XS):KE$=RIGHT$(XS$,3):KE=VAL(KE$)
10085 XS=VAL(LEFT$(XS$,LEN(XS$)-4))
10090 GOSUB10110
10095 RETURN
10110 REM FORM O/P$
10115 GOSUB10145
10120 IFABS(XS)>=10THENGOSUB10175
10125 GOSUB10200
10130 GOSUB10225
10135 RETURN
10145 REM ROUNDOFF MANTISSA
```

```
10155 XR=5:FORI5=1TONS:XR=XR/10:NEXTI5
10160 XS=XS+XR*SGN(XS)
10165 RETURN
10175 REM NORMALISE MANTISSA
10180 IF ABS(XS)<1THENXS=XS*10:KE=KE-1:GOSUB10180
10185 IF ABS(XS)>=10THENXS=XS/10:KE=KE+1:GOSUB10185
10190 RETURN
10200 REM FORM EXPONENT$
10205 S$="+":IFKE<0THENS$="-"
10210 KE$=S$+RIGHT$("0"+MID$(STR$(KE),2),2)
10215 RETURN
10225 REM FORM (MANT+EXP)$
10230 X1$=LEFT$(STR$(XS),NS+2)
10235 XS$=X1$+LEFT$(B0$,NS+2-LEN(X1$))
10240 IFXS=INT(XS)THEN XS$=X1$+"."+LEFT$(B0$,NS-1)
10245 XS$=LEFT$(BL$,FW-NS-6)+XS$+"E"+KE$
10250 RETURN
20000 REM  **  FORMAT NUMERICAL OUTPUT            **
20020 FW=12:NS=3
20040 BL$="                   "
20050 XS$=STR$(XS):XE$="     ":IFLEN(XS$)<4THENXS$=XS$+"    "
20060 IFABS(XS)>=10↑(8-NS)THENXX=XS:GOTO20080
20070 XX=XS+.5*SGN(XS)/10↑NS
20080 IFMID$(XS$,LEN(XS$)-3,1)="E"THENGOSUB20180
20090 XX$=STR$(XX)
20100 FORJ5=1TOLEN(XX$)
20110 IFMID$(XX$,J5,1)="."THENDP=J5:GOTO20130
20120 NEXTJ5:DP=J5:XX$=XX$+".0000000"
20130 XS$=LEFT$(XX$,DP+NS)+XE$
20140 LX=LEN(XS$):IFLX>FWTHENXS$=LEFT$(XS$,FW):GOTO20160
20150 XS$=LEFT$(BL$,FW-LX)+XS$
20160 RETURN
20180 XE$=RIGHT$(XS$,4):XR=VAL(RIGHT$(XS$,2))
20190 XX=VAL(LEFT$(XS$,LEN(XS$)-4))+.5*SGN(XS)/10↑NS
20200 RETURN
50000 REM ** THE FOLLOWING STATEMENTS
50001 REM    MUST BE TAILORED TO THE
50002 REM     PARTICULAR MACHINE IN USE          **
50005 PRINT#5,P$:RETURN :REM ** PRINTLINE ON PRINTER **
50010 PRINT#5:RETURN    :REM ** 1 LINEFEED ON PRINTER **
50015 REM
50020 FOR KP = 1 TO IP  :REM IP
50021 PRINT#5            :REM   LINEFEEDS
50022 NEXT KP           :REM     ON
50023 RETURN            :REM      PRINTER *
50025 REM
51000 OPEN 5,4:RETURN   :REM ** OPEN CHANNEL TO PRINTER  **
52000 CLOSE 5:RETURN    :REM ** CLOSE CHANNEL TO PRINTER **
53000 REM ** THE FOLLOWING Y$,N$,L4$,AK$
53001 REM    ARE USED WITH 'INPUT' STATEMENTS -
53002 REM      SET THEM ALL EQUAL TO "" IF
53003 REM      L4$ NOT POSSIBLE ON YOUR MACHINE     **
53010 Y$="Y ":N$="N ":AK$="* "
53020 L4$="▉▉▉▉▉":RETURN :REM ** L4$ = 4 'CURSOR-LEFTS' **
```

3.2.4 *Example of plane frame analysis using **PLFRAME***

(a)   **The structure.** A propped cantilever carrying a uniformly distributed load.

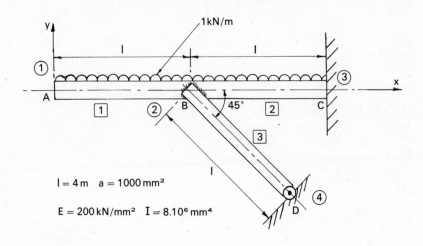

$l = 4$ m $a = 1000$ mm²

$E = 200$ kN/mm² $I = 8.10^6$ mm⁴

(b)   *The data.*

```
4000 DATA 4,3
4010 DATA 1,0,0,0,0,0,0,-2,-1333
4020 DATA 2,4000,0,0,0,0,0,-4,0
4030 DATA 3,8000,0,1,1,1,0,-2,1333
4040 DATA 4,6828.4,-2828.4,1,1,1,0,0,0
4050 DATA 1,1,2,0,8E6,1E3,200
4060 DATA 2,2,3,0,8E6,1E3,200
4070 DATA 3,2,4,2,8E6,1E3,200
```

(c) *The solution.*

```
PROGRAM PLFRAME


                        *** DATA INPUT ***


NUMBER OF NODES   =   4
NUMBER OF MEMBERS =   3


* NODAL DATA *

NODE      COORDINATES           R  E  S  T  R  A  I  N  T  S
 NO     X           Y           U              V           A N G L E
 1     0.000       0.000        0              0              0
 2     4000.000    0.000        0              0              0
 3     8000.000    0.000        1              1              1
 4     6828.400   -2828.400     1              1              1


* MEMBER DATA *

MEMBER END NODES   MEMBER   2ND MOMT    SECTN      ELASTIC
  NO    I    J      TYPE     OF AREA     AREA       MODULUS
  1     1    2       0      8.00E+06    1.00E+03    2.00E+02
  2     2    3       0      8.00E+06    1.00E+03    2.00E+02
  3     2    4       2      8.00E+06    1.00E+03    2.00E+02

* APPLIED NODAL FORCES *

NODE     F  O  R  C  E  S        MOMENT
 NO      X           Y             M
 1      0.000      -2.000      -1333.000
 2      0.000      -4.000          0.000
 3      0.000      -2.000       1333.000
 4      0.000       0.000          0.000
```

D

```
                    *** OUTPUT OF RESULTS ***

* VECTOR OF ALL DISPLACEMENTS *

NODE           DISPLACEMENTS                ROTATION
NO        U                  V              ANGLE

1        -.168            -30.621           9.202E-03
2        -.168             -.484            2.534E-03
3        0.000             0.000            0.000
4        0.000             0.000            0.000

* MEMBER END FORCES W.R.T GLOBAL COORDS *

MEMBER NODE          F O R C E S              MOMENT
  NO   NO        X                  Y              M

   1    1      0.000            -2.000         -1333.000
   1    2      0.000             2.000         -6667.000

   2    2     -8.402             1.375          3764.339
   2    3      8.402            -1.375          1737.099

   3    2      8.402            -7.375          2902.661
   3    4     -8.402             7.375             0.000


* NODAL LOADS W.R.T. GLOBAL COORDS *

NODE        F O R C E S              MOMENT
NO       X              Y              M
 1      0.000         -2.000        -1333.000

 2     3.147E-10      -4.000         4.143E-06

 3      8.402         -1.375         1737.099

 4     -8.402          7.375            0.000

         **** END OF RUN OF PROGRAM PLFRAME ****
```

Appendix 3.1 PLFRAME: (Computer program for plane frame analysis): program summary and data sheet

1. *Introduction*

From data read in regarding joints and members the program assembles a stiffness matrix for each member about its own axis and then a rotation matrix. A triple product of these two matrices transforms the member stiffness matrix to being about the structure origin. If the joints at the end of each member are denoted i and j, the original equations relating end actions to end displacements and rotations can be written as shown on p.92.

The transformed member stiffness matrices are successively added into the global stiffness matrix. The load vector is formed from information on nodal load data cards. Member loads should be converted to equivalent joint loads.

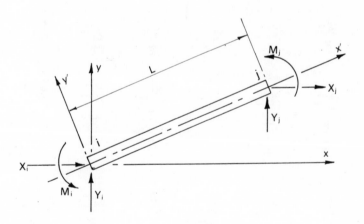

$$
[\mathbf{K}] = \frac{EA}{L}
\begin{array}{cccccc}
u_i & v_i & \theta_i & u_j & v_j & \theta_j \\
l^2+m^2\dfrac{12k^2}{L^2} & lm-lm\dfrac{12k^2}{L^2} & -m\dfrac{6k^2}{L} & -l^2-m^2\dfrac{12k^2}{L^2} & -lm+lm\dfrac{12k^2}{L^2} & -m\dfrac{6k^2}{L} \\[2.2ex]
lm-lm\dfrac{12k^2}{L^2} & m^2+l^2\dfrac{12k^2}{L^2} & l\dfrac{6k^2}{L} & -lm+lm\dfrac{12k^2}{L^2} & -m^2-l^2\dfrac{12k^2}{L^2} & l\dfrac{6k^2}{L} \\[2.2ex]
-m\dfrac{6k^2}{L} & l\dfrac{6k^2}{L} & 4k^2 & m\dfrac{6k^2}{L} & -l\dfrac{6k^2}{L} & 2k^2 \\[2.2ex]
-l^2-m^2\dfrac{12k^2}{L^2} & -lm+lm\dfrac{12k^2}{L^2} & m\dfrac{6k^2}{L} & l^2+m^2\dfrac{12k^2}{L^2} & lm-lm\dfrac{12k^2}{L^2} & m\dfrac{6k^2}{L} \\[2.2ex]
-lm+lm\dfrac{12k^2}{L^2} & -m^2-l^2\dfrac{12k^2}{L^2} & -l\dfrac{6k^2}{L} & lm-lm\dfrac{12k^2}{L^2} & m^2+l^2\dfrac{12k^2}{L^2} & -l\dfrac{6k^2}{L} \\[2.2ex]
-m\dfrac{6k^2}{L} & l\dfrac{6k^2}{L} & 2k^2 & m\dfrac{6k^2}{L} & -l\dfrac{6k^2}{L} & 4k^2
\end{array}
$$

$$(3.27)$$

Preparation of input data for this program should be accomplished in the following sequence:

(1) Sketch the structure and number the joints and members.
(2) Establish a reference coordinate system and label joints with proper coordinate values.
(3) Define the different load cases to be considered.
(4) Fill out datasheet.

2. Datasheet for **PLFRAME**

Note: units must be consistent

Structure data. Number of nodes (NN ≤ 10), number of members (NM ≤ 10)

	,	

Nodal data. Node number, its coordinates, its restraints (=1 for zero displacement of slope, else = 0) and applied loads for each node starting with ① and ending with NN.

Node No	Coordinates		Restraints on			Applied nodal loads		
	X	Y	u	v	θ	P_x	P_y	M_z
,	,	,	,	,	,	,	,	
,	,	,	,	,	,	,	,	
,	,	,	,	,	,	,	,	
,	,	,	,	,	,	,	,	
,	,	,	,	,	,	,	,	
,	,	,	,	,	,	,	,	
,	,	,	,	,	,	,	,	
,	,	,	,	,	,	,	,	
,	,	,	,	,	,	,	,	
,	,	,	,	,	,	,	,	

Element data. Element number, its end nodes and type (=0 for fixed/fixed, 1 for pinned/fixed, 2 for fixed/pinned and 3 for pinned/pinned), its second moment of area and cross-section area, and its modulus of elasticity for each element starting with ① and ending with NE.

Element No.	Node Numbers		Type	Second Moment of Area	Area	Modulus of Elasticity
,	,	,	,	,	,	
,	,	,	,	,	,	
,	,	,	,	,	,	
,	,	,	,	,	,	
,	,	,	,	,	,	
,	,	,	,	,	,	
,	,	,	,	,	,	
,	,	,	,	,	,	

Notes: (a) This is merely a sample blank datasheet for up to 10 nodes and 8 elements. (b) The data should be typed into the program in DATA statements between line numbers 4000 and 10000.

4 FEP: plane stress/plane strain finite element analysis

4.0 Introduction

The finite element approach to continuum problems developed as a direct extension of the approach used in pin-jointed plane frames and subsequently beam elements analysis.

To outline the philosophy consider the thin triangular plate ABC, fixed along its edge BC and loaded at D and A along edge CDA in its plane (Fig. 4.1).

It is required to find the displacements under the loads and the stresses in the plate.

The steps in a finite element analysis are as follows:

(a) Divide the plate up into a number of elements of finite dimensions, for example triangles, joined together at their corners or nodes, so that the corners of adjacent elements have common displacements. This process is referred to as discretization. An example of a very coarse division is shown in the figure: the more elements taken, the better would be the results.

(b) *Assume* a state of strain in each element and express it in terms of the nodal displacements. Then use the stress/strain relations to obtain the stresses in the element and find a set of *fictitious* nodal forces on each element, which would be in equilibrium with the internal stresses. These fictitious nodal forces are thus expressed in terms of the nodal displacements. The most common assumption for a state of strain in an element is that of uniform strain, i.e. the same state of strain at all points in an element but differnt from element to element.

(c) Apply the conditions for equilibrium of each node under the forces applied to it from adjacent elements, together with any external forces applied at the node. This gives a set of equations for the unknown nodal displacements. For the crude subdivisions into triangular elements shown in the figure, there would be 6 unknown nodal displacements; u_A, v_A, u_D, v_D, u_E, v_E and 6 equations of equilibrium for the 3 nodes A, D and E ($u_B = v_B = u_C = v_C = 0$). When all the nodal displacements have been determined, the stresses in each element can be evaluated.

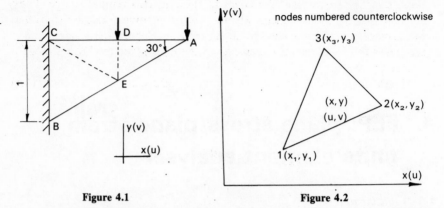

Figure 4.1   Figure 4.2

This procedure seems rather suspect at first sight – assuming states of strain in the elements, introducing fictitious nodal forces etc. – but it can be shown that if correctly used, it is merely an applicaton of the theorem of minimum potential energy. The formal framework of the minimum potential energy theorem defines the restrictions on what deformation states may be assumed.

The development of the equations has the same aim as that in the previous two chapters; stiffness coefficients must be established relating the nodal displacements to nodal forces within the element, the full matrix of coefficients forming the stiffness matrix. However, whereas within the assumptions made in the moment curvature relation in section 3.1.1, a more rigorous derivation of the beam elements is possible, no such rigour is possible for the continuum. Assumptions must be made about the displacement distribution within an element, usually described as some polynomial, the higher the order of which, the more accurate the description of the distribution produced. Only the simplest polynomial will be used here and thus the simplest of elements is incorporated into program **FEP**.

4.1 Development of equations

4.1.1 *The stiffness coefficients for a general triangular plane stress /plain strain element*

The simplest assumption for the deformation within an element is a uniform state of strain throughout. This may be achieved by assuming the displacements u and v to be given by linear functions of x and y, the global coordinates.
Therefore,

$$u = \alpha_1 + \alpha_2 x + \alpha_3 y$$
$$v = \beta_1 + \beta_2 x + \beta_3 y \qquad (4.1)$$

where $\alpha_1, \alpha_2, \beta_1, \beta_2$ and β_3 are constants.

Note: this also means that the edges of the elements remain straight lines. Furthermore, since two adjacent elements have common nodes, it means that the displacements u and v are continuous from one element to the next – one of the conditions for elements to conform to the rules of the minimum potential energy theorem.

The strains in the element corresponding to the displacement functions (4.1) are:

$$\epsilon_x = \frac{\partial u}{\partial x} = \alpha_2$$

$$\epsilon_y = \frac{\partial v}{\partial y} = \beta_3$$

$$\gamma_{xy} = \frac{\partial u}{\partial y} + \frac{\partial v}{\partial x} = \alpha_3 + \beta_2$$

$$(4.2)$$

It can be clearly seen that the strains are constant, thus the element is called a *constant strain triangle,* or CST.

The constants and hence the strains ϵ_x, ϵ_y, γ_{xy} can be obtained in terms of the nodal displacements $u_1, v_1, u_2, v_2, u_3, v_3$ by applying the equations (4.1) to each node

$$\left. \begin{array}{l} u_1 = \alpha_1 + \alpha_2 x_1 + \alpha_3 y_1 \\ v_1 = \beta_1 + \beta_2 x_1 + \beta_3 y_1 \\ u_2 = \alpha_1 + \alpha_2 x_2 + \alpha_3 y_2 \\ v_2 = \beta_1 + \beta_2 x_2 + \beta_3 y_2 \\ u_3 = \alpha_1 + \alpha_2 x_3 + \alpha_3 y_3 \\ v_3 = \beta_1 + \beta_2 x_3 + \beta_3 y_3 \end{array} \right\} \quad (4.3)$$

or

$$\{U\} = \begin{bmatrix} u_1 \\ v_1 \\ u_2 \\ v_2 \\ u_3 \\ v_3 \end{bmatrix} = \begin{bmatrix} 1 & x_1 & y_1 & 0 & 0 & 0 \\ 0 & 0 & 0 & 1 & x_1 & y_1 \\ 1 & x_2 & y_2 & 0 & 0 & 0 \\ 0 & 0 & 0 & 1 & x_2 & y_2 \\ 1 & x_3 & y_3 & 0 & 0 & 0 \\ 0 & 0 & 0 & 1 & x_3 & y_3 \end{bmatrix} \begin{bmatrix} \alpha_1 \\ \alpha_2 \\ \alpha_3 \\ \beta_1 \\ \beta_2 \\ \beta_3 \end{bmatrix} = [A]\{\alpha\}$$

$$(4.4)$$

Similarly, (4.2) can be rewritten as

$$\{\epsilon\} = \begin{bmatrix} \epsilon_x \\ \epsilon_y \\ \gamma_{xy} \end{bmatrix} = \begin{bmatrix} 0 & 1 & 0 & 0 & 0 & 0 \\ 0 & 0 & 0 & 0 & 0 & 1 \\ 0 & 0 & 1 & 0 & 1 & 0 \end{bmatrix} \{\alpha\} = [B_\alpha]\{\alpha\}$$

$$(4.5)$$

The constants $\{\alpha\}$ are of no further use, however, and should be eliminated such that the strain can be expressed in terms of nodal displacements $\{U\}$.

Before this is done the area of the element is calculated in terms of the nodal coordinates (Fig. 4.3).

Area of triangle 123 = area 1354 + area 2356 – area 1462

$$= \tfrac{1}{2}\left[(y_1 + y_3)(x_3 - x_1) + (y_3 + y_2)(x_2 - x_3) - (y_1 + y_2)(x_2 - x_1)\right]$$

Therefore, area of element 'A' $= \tfrac{1}{2}\left[y_1x_3 - y_3x_1 + y_3x_2 - y_2x_3 + y_2x_1 - y_1x_2\right]$
$= \tfrac{1}{2}\lambda$, say.

Or $\lambda = 2.A.$

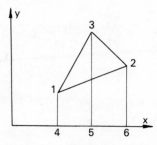

Figure 4.3

Note the area A is positive *if* the element is numbered *anticlockwise*.

Returning to equations (4.4), the constants α_2, α_3, β_2, β_3 can be solved in terms of the other variables such that

$$
\left.
\begin{aligned}
\alpha_2 &= \tfrac{1}{2}A\left[(y_2 - y_3)u_1 + (y_3 - y_1)u_2 + (y_1 - y_2)u_3\right] \\
\alpha_3 &= -\tfrac{1}{2}A\left[(x_2 - x_3)u_1 + (x_3 - x_1)u_2 + (x_1 - x_2)u_3\right] \\
\beta_2 &= \tfrac{1}{2}A\left[(y_2 - y_3)v_1 + (y_3 - y_1)v_2 + (y_1 - y_2)v_3\right] \\
\beta_3 &= -\tfrac{1}{2}A\left[(x_2 - x_3)v_1 + (x_3 - x_1)v_2 + (x_1 - x_2)v_3\right]
\end{aligned}
\right\} \quad (4.6)
$$

Substitution of equations (4.6) into (4.2) now gives the element strains in terms of the nodal displacements.

$$\epsilon_x = \tfrac{1}{2}A\left[(y_2 - y_3)u_1 + (y_3 - y_1)u_2 + (y_1 - y_2)u_3\right]$$

$$\epsilon_y = \tfrac{1}{2}A\left[(x_3 - x_2)v_1 + (x_1 - x_3)v_2 + (x_2 - x_1)v_3\right]$$

$$\gamma_{xy} = \tfrac{1}{2}A\left[(x_3 - x_2)u_1 + (x_1 - x_3)u_2 + (x_2 - x_1)u_3 + (y_2 - y_3)v_1 + (y_3 - y_1)v_2 \right.$$
$$\left. + (y_1 - y_2)v_3\right]$$

or in matrix terms,

$$\{\epsilon\} = \frac{1}{2A} \begin{bmatrix} b_1 & 0 & b_2 & 0 & b_3 & 0 \\ 0 & a_1 & 0 & a_2 & 0 & a_3 \\ a_1 & b_1 & a_2 & b_2 & a_3 & b_3 \end{bmatrix} \{U\} = [B]\{U\}$$

(4.7)

where

$$a_1 = x_3 - x_2 \qquad \text{and } b_1 = y_2 - y_3$$
$$a_2 = x_1 - x_3 \qquad \qquad b_2 = y_3 - y_1$$
$$a_3 = x_2 - x_1 \qquad \qquad b_3 = y_1 - y_2$$

In order to determine the stresses which correspond to these above strains, it is necessary to decide on specific stress/strain relations. The constitutive equations are for an isotropic linear elastic material with elastic modulus E, shear modulus G and Poisson's ratio v. The two limiting plane states are considered – plane stress and plane strain.

In *plane stress,* the through thickness *stress* $\sigma_2 = 0$ and through thickness displacements are unrestrained.

Thus,

$$\sigma_x = \frac{E}{1 - v^2} (\epsilon_x + v\epsilon_y)$$

$$\sigma_y = \frac{E}{1 - v^2} (\epsilon_y + v\epsilon_x)$$

$$\tau_{xy} = G.\gamma_{xy}$$

or in matrix form,

$$\{\sigma\} = \begin{bmatrix} \sigma_x \\ \sigma_y \\ \tau_{xy} \end{bmatrix} = \begin{bmatrix} \dfrac{E}{1 - v^2} & \dfrac{vE}{1 - v^2} & 0 \\ \dfrac{vE}{1 - v^2} & \dfrac{E}{1 - v^2} & 0 \\ 0 & 0 & \dfrac{E}{2(1 + v)} \end{bmatrix} \{\epsilon\} = [D]\{\epsilon\}$$

(4.8)

Note that $E = 2G(1 + v)$.

The coefficients of the $[\mathbf{D}]$ matrix for *plane stress* are

$$\left.\begin{aligned}
d_{11} = d_{22} &= \frac{E}{1-v^2} \\[2mm]
d_{21} = d_{12} &= \frac{vE}{1-v^2} \\[2mm]
d_{33} = G &= \frac{E}{2(1+v)}
\end{aligned}\right\} \quad (4.9)$$

In *plane strain,* $\epsilon_z = 0$ and $\sigma_z = v\,(\sigma_x + \sigma_y)$ and there is zero through thickness displacement.

Since σ_z can be derived from σ_x and σ_y it is not included in the vector $\{\sigma\}$. The coefficients in $[\mathbf{D}]$ for *plane strain* are

$$d_{11} = d_{22} = \frac{E(1-v)}{(1+v)\,(1-2v)}$$

$$d_{12} = d_{21} = \frac{vE}{(1+v)\,(1-2v)}$$

$$d_{33} = \frac{E}{2(1+v)}$$

or

$$[\mathbf{D}] = \begin{bmatrix}
\dfrac{E(1-v)}{(1+v)\,(1-2v)} & \dfrac{vE}{(1+v)\,(1-2v)} & 0 \\[4mm]
\dfrac{vE}{(1+v)\,(1-2v)} & \dfrac{E(1-v)}{(1+v)\,(1-2v)} & 0 \\[4mm]
0 & 0 & \dfrac{E}{2(1+v)}
\end{bmatrix}$$

$$(4.10)$$

Note that the inclusion of the $(1-2v)$ term in the denominators means that there is no solution possible for the limiting condition of $v = 0.5$, where there is no matrix volume change in the material during deformation.

Substituting $\{\epsilon\} = [\mathbf{B}]\,\{\mathbf{U}\}$ (4.7) into $\{\sigma\} = [\mathbf{D}]\,\{\epsilon\}$ (4.8) leads to

$$\{\sigma\} = [\mathbf{D}]\,\{\epsilon\} = [\mathbf{D}]\,[\mathbf{B}]\,\{\mathbf{U}\} \tag{4.11}$$

which means that the element stresses are now written in terms of the nodal displacements. So:

$$\left.\begin{array}{l} \sigma_x = \dfrac{1}{2A}\left\{ \begin{array}{l} d_{11}\,(y_2 - y_3)u_1 + d_{11}\,(y_3 - y_1)u_2 + d_{11}\,(y_1 - y_2)u_3 \\ +d_{12}\,(x_3 - x_2)v_1 + d_{12}\,(x_1 - x_3)v_2 + d_{12}\,(x_2 - x_1)v_3 \end{array}\right. \\[2em] \sigma_y = \dfrac{1}{2A}\left\{ \begin{array}{l} d_{12}\,(y_2 - y_3)u_1 + d_{12}\,(y_3 - y_1)u_2 + d_{12}\,(y_1 - y_2)u_3 \\ d_{11}\,(x_3 - x_2)v_1 + d_{11}\,(x_1 - x_3)v_2 + d_{11}\,(x_2 - x_1)v_3 \end{array}\right. \\[2em] \tau_{xy} = \dfrac{1}{2A}\left\{ \begin{array}{l} d_{33}\,(x_3 - x_2)u_1 + d_{33}\,(x_1 - x_3)u_2 + d_{33}\,(x_2 - x_1)u_3 \\ +d_{33}\,(y_2 - y_3)v_1 + d_{33}\,(y_3 - y_1)v_2 + d_{33}\,(y_1 - y_2)v_3 \end{array}\right. \end{array}\right\} \tag{4.12}$$

There remains the last step – to relate the nodal forces corresponding to the above stresses to the nodal displacements and thus determine the stiffness coefficients of the stiffness matrix. To obtain the set of nodal forces in equilibrium with the stresses in the element (4.12), use is made of the *equation of virtual work*.

For plane stress conditions in a continuum, and with point forces only on the boundary (Fig. 4.4), the virtual work equation is

$$\int_v (\sigma_x \epsilon_x + \sigma_y \epsilon_y + \tau_{xy} \gamma_{xy}) dV$$

$$- X_1 u_1 - X_2 u_2 - X_3 u_3 - Y_1 v_1$$

$$- Y_2 v_2 - Y_3 v_3 = 0 \tag{4.13}$$

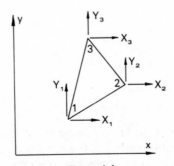

Figure 4.4

or in matrix form

$$\overbrace{\iiint \{\epsilon\}^{T} \{\sigma\} dxdyxz}^{\text{equilibrium set}} \underbrace{- \{U\}^{T} \{P\}}_{\text{deformation set}} = O$$

where $\{P\} = \begin{bmatrix} X_1 \\ Y_1 \\ X_2 \\ Y_2 \\ X_3 \\ Y_3 \end{bmatrix}$ and $\{\epsilon\}^{T}$ and $\{U\}^{T}$ indicate the transpose of the column vector to a row vector, viz. $\begin{bmatrix} \epsilon_x & \epsilon_y & \gamma_{xy} \end{bmatrix}$

and $\begin{bmatrix} u_1 & v_1 & y_2 & v_2 & u_3 & v_3 \end{bmatrix}$.

Now in equation (4.13), the stress $\{\sigma\}$ and the nodal (external) forces $\{P\}$ form *any* equilibrium set and the strains $\{\epsilon\}$ with the nodal (boundary) displacements $\{U\}$ form *any* compatible deformation or geometry set for this particular element. The 'magic' of the virtual work equation is that for a particular problem, the equilibrium set corresponding to one loading can be multiplied with the deformation set corresponding to another loading and the result will still be zero. This property is used to great effect to determine the stiffness coefficients by selecting deformation sets $\{U'\}$ corresponding to some unusual loading $\{P'\}$ which greatly simplifies finding the relations between $\{\sigma\}$ (written in terms of $\{U\}$ (4.12), and $\{P\}$.

For X_1 the equilibrium set is

$$\sigma_x, \sigma_y, \tau_{xy} \text{ with } X_1, Y_1, X_2, Y_2, X_3, Y_3$$

and the geometry or deformation set is

$$u'_1, \neq 0, v'_1 = u'_2 = v'_2 = u'_3 = v'_3 = 0$$

produced by some unusual loading $(X'_1, Y'_1$ etc.), which in fact need not be determined here.

From (4.6) $\epsilon'_x = \dfrac{1}{2A} (y_2 - y_3) u'_1$

$$\epsilon'_y = 0$$

$$\gamma'_{xy} = \dfrac{1}{2A} \left[(x_3 - x_2) u'_1 \right]$$

Equation (4.13) thus gives

$$\int_V \left[\sigma_x \epsilon'_x + \sigma_y \epsilon'_y + \tau_{xy} \gamma_{xy}' \right] dV - X_1 u'_1 = 0$$

or

$$\int_V \left[\sigma_x \frac{1}{2A} (y_2 - y_3) u'_1 + 0 + \tau_{xy} \frac{1}{2A} (x_3 - x_2) y'_1 \right] dV - X_1 u'_1 = 0$$

leading to $X_1 = \frac{1}{2A} \int_V \left[\sigma_x (y_2 - y_3) + \tau_{xy} (x_3 - x_2) \right] dV.$

Since in this CST element σ_x and τ_{xy} are constant through the element,

$$X_1 = \frac{V}{2A} \left[\sigma_x (y_2 - y_3) + \tau_{xy} (x_3 - x_2) \right].$$

With element volume $V = A.t$ where t is the element thickness, and using equations (4.12), X_1 can now be written in terms of the nodal displacements, thus giving a set of stiffness coefficients.

$$X_1 = \frac{t}{4A} \left\{ \begin{array}{l} u_1 \left[d_{11}(y_2 - y_3)^2 \quad\quad + d_{33} (x_3 - x_2)^2 \right] \\ +v_1 \left[d_{12} (x_3 - x_2)(y_2 - y_3) \quad + d_{33} (y_2 - y_3)(x_3 - x_2) \right] \\ +u_2 \left[d_{11} (y_3 - y_1)(y_2 - y_3) \quad + d_{33} (x_1 - x_3)(x_3 - x_2) \right] \\ +v_2 \left[d_{12} (x_1 - x_3)(y_2 - y_3) \quad + d_{33} (y_3 - y_1)(x_3 - x_2) \right] \\ +u_3 \left[d_{11} (y_1 - y_2)(y_2 - y_3) \quad + d_{33} (x_2 - x_1)(x_3 - x_2) \right] \\ +v_3 \left[d_{12} (x_2 - x_1)(y_2 - y_3) \quad + d_{33} (y_1 - y_2)(x_3 - x_2) \right] \end{array} \right\}$$

The other nodal forces are obtained in a similar way. For example, to find Y_2, take the equilibrium set as before but with the geometry set $v'_2 \neq 0$

and $\quad\quad\quad v'_1 = u'_1 = u'_2 = u'_3 = v'_3 = 0,$

giving $\quad\quad \epsilon'_x = 0, \epsilon'_y = \frac{1}{2A}(x_1 - x_3)v'_2$ and $\gamma'_{xy} = \frac{1}{2A}(y_3 - y_1)v'_2$

and so on.

The complete set of stiffness coefficients (4.14) for the uniform strain triangle (or CST) under isotropic linear elastic plane stress conditions is given below (Table 4.1). For plane strain the matrix is identical with the substitution of the plane strain d's from (4.10) instead of (4.9). The matrix is symmetrical.

It may be easier to think in terms of the matrix formulation. From (4.14).

$$\iiint\limits_V \{\epsilon\}^T \{\sigma\} \, dV = \{U\}^T \{P\}$$

but with (4.7) and (4.8)

$$\iiint\limits_V \{U\}^T [B]^T [D] [B] \{U\} \, dV = \{U\}^T \{P\}$$

and since stress and strain are constant over the element and $V = At$

$$\{U\}^T [B]^T [D] [B] \{U\} V = \{U\}^T \{P\} \quad \text{or} \quad [K] \{U\} = \{P\}$$

where the stiffness matrix $[K]$ is the triple product

$$[K] = t.A. [B]^T [D] [B]$$

and $[B]$ is given by equation (4.7) and $[D]$ from either (4.9) or (4.10).

Usually in computer programs the explicit form of $\left[K\right]$ (4.14), is not stored but built up from the matrices $\left[B\right]$ and $\left[D\right]$ and the product $\left[D\right] \left[B\right]$ stored in order that the element stresses can be immediately found (from equation (4.11)) once the nodal displacements have been found.

4.2 Illustrative example

In order to illustrate longhand the application of finite elements to a particular problem, consider the structure shown in Fig. 4.1 and again below in Fig. 4.5.

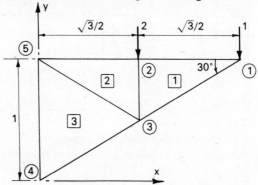

Figure 4.5

Number the nodes ① to ⑤ as shown with the origin of coordinates at node ④. The elements are ① numbered anti-clockwise ①,②,③, ② as ⑤,③,② and ③ as ⑤,④,③. Assume the cantilever to have unit thickness and be in plane stress with $E = 1$ and $v = 0.3$. Consider each element in turn and establish the stiffness matrix.

Element ①:

$x_1 = \sqrt{3}$	$y_1 = 1$
$x_2 = \sqrt{3}/2$	$y_2 = 1$
$x_3 = \sqrt{3}/2$	$y_3 = \frac{1}{2}$

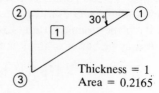

Thickness = 1
Area = 0.2165

$$a_1 = x_3 - x_2 = 0 \qquad b_1 = y_2 - y_3 = \tfrac{1}{2}$$
$$a_2 = x_1 - x_3 = \sqrt{3}/2 \qquad b_2 = y_3 - y_1 = -\tfrac{1}{2}$$
$$a_3 = x_2 - x_1 = -\sqrt{3}/2 \qquad b_3 = y_1 - y_2 = 0$$

Table 4.1 Equation (4.14) for a constant strain triangle

$$\frac{t}{4A}
\begin{bmatrix}
d_{11}(y_2-y_3)^2 & d_{12}(x_3-x_2)(y_2-y_3) & d_{11}(y_2-y_3)(y_3-y_1) & d_{12}(x_1-x_3)(y_2-y_3) & d_{11}(y_2-y_3)(y_1-y_2) & d_{12}(x_2-x_1)(y_2-y_3) \\
+\,d_{33}(x_3-x_2)^2 & +\,d_{33}(y_2-y_3)(x_3-x_2) & +\,d_{33}(x_3-x_2)(x_1-x_3) & +\,d_{33}(y_3-y_1)(x_3-x_2) & +\,d_{33}(x_3-x_2)(x_2-x_1) & +\,d_{33}(y_1-y_2)(x_3-x_2) \\[6pt]
 & d_{11}(x_3-x_2)^2 & d_{12}(y_3-y_1)(x_3-x_2) & d_{11}(x_1-x_3)(x_3-x_2) & d_{12}(y_1-y_2)(x_3-x_2) & d_{11}(x_2-x_1)(x_3-x_2) \\
 & +\,d_{33}(y_2-y_3)^2 & +\,d_{33}(x_1-x_3)(y_2-y_3) & +\,d_{33}(y_3-y_1)(y_2-y_3) & +\,d_{33}(x_2-x_1)(y_2-y_3) & +\,d_{33}(y_1-y_2)(y_2-y_3) \\[6pt]
 & & d_{11}(y_3-y_1)^2 & d_{12}(x_1-x_3)(y_3-y_1) & d_{11}(y_3-y_1)(y_1-y_2) & d_{12}(x_2-x_1)(y_3-y_1) \\
 & & +\,d_{33}(x_1-x_3)^2 & +\,d_{33}(y_3-y_1)(x_1-x_3) & +\,d_{33}(x_1-x_3)(x_2-x_1) & +\,d_{33}(y_1-y_2)(x_1-x_3) \\[6pt]
 & & & d_{11}(x_1-x_3)^2 & d_{12}(y_1-y_2)(x_1-x_3) & d_{11}(x_2-x_1)(x_1-x_3) \\
 & & & +\,d_{33}(y_3-y_1)^2 & +\,d_{33}(x_2-x_1)(y_3-y_1) & +\,d_{33}(y_1-y_2)(y_3-y_1) \\[6pt]
 & & \text{SYMMETRICAL} & & d_{11}(y_1-y_2)^2 & d_{12}(x_2-x_1)(y_1-y_2) \\
 & & & & +\,d_{33}(x_2-x_1)^2 & +\,d_{33}(y_1-y_2)(x_2-x_1) \\[6pt]
 & & & & & d_{11}(x_2-x_1)^2 \\
 & & & & & +\,d_{33}(y_1-y_2)^2
\end{bmatrix}
\begin{Bmatrix} u_1 \\ v_1 \\ u_2 \\ v_2 \\ u_3 \\ v_3 \end{Bmatrix}
=
\begin{Bmatrix} X_1 \\ Y_1 \\ X_2 \\ Y_2 \\ X_3 \\ Y_3 \end{Bmatrix}$$

Table 4.2

$$\begin{bmatrix} {}_1X_1 \\ {}_1Y_1 \\ {}_1X_2 \\ {}_1Y_2 \\ {}_1X_3 \\ {}_1Y_3 \end{bmatrix} = \left(\frac{1}{4 \cdot 0.2165}\right)$$

	u_1	v_1	u_2	v_2	u_3	v_3
${}_1X_1$	$d_{11}.(\tfrac{1}{2})^2 + d_{33}.(0)^2$	$d_{12}.0.(\tfrac{1}{2}) + d_{33}.(\tfrac{1}{2}).0$	$d_{11}.(-\tfrac{1}{2}).(\tfrac{1}{2}) + d_{33}.(\sqrt{3}/2).0$	$d_{12}.(\sqrt{3}/2).(\tfrac{1}{2}) + d_{33}.(-\tfrac{1}{2}).0$	$d_{11}.0.(\tfrac{1}{2}) + d_{33}.0.0$	$d_{12}.(-\sqrt{3}/2).(\tfrac{1}{2}) + d_{33}.0.0$
${}_1Y_1$		$d_{11}.0^2 + d_{33}.(\tfrac{1}{2})^2$	$d_{12}.(-\tfrac{1}{2}).0 + d_{33}.(\sqrt{3}/2).(\tfrac{1}{2})$	$d_{11}.\sqrt{3}/2.0 + d_{33}.(-\tfrac{1}{2}).(\tfrac{1}{2})$	$d_{12}.0.0 + d_{33}.0.(\tfrac{1}{2})$	$d_{11}.(-\sqrt{3}/2).0 + d_{33}.0.(\tfrac{1}{2})$
${}_1X_2$			$d_{11}.(-\tfrac{1}{2})^2 + d_{33}.(\sqrt{3}/2)^2$	$d_{12}.(\sqrt{3}/2).(-\tfrac{1}{2}) + d_{33}.(-\tfrac{1}{2}).(\sqrt{3}/2)$	$d_{11}.0.(-\tfrac{1}{2}) + d_{33}.0.(\sqrt{3}/2)$	$d_{12}.(-\sqrt{3}/2).(-\tfrac{1}{2}) + d_{33}.0.(\sqrt{3}/2)$
${}_1Y_2$				$d_{11}.(\sqrt{3}/2)^2 + d_{33}.(-\tfrac{1}{2})^2$	$d_{12}.0.(\sqrt{3}/2) + d_{33}.0.(-\tfrac{1}{2})$	$d_{11}.(-\sqrt{3}/2).(\sqrt{3}/2) + d_{33}.0.(-\tfrac{1}{2})$
${}_1X_3$					$d_{11}.0^2 + d_{33}.0^2$	$d_{12}.(-\sqrt{3}/2).0 + d_{33}.0.(-\sqrt{3}/2)$
${}_1Y_3$						$d_{11}.(-\sqrt{3}/2)^2 + d_{33}.0^2$

SYMMETRICAL

Using (4.14) gives Table 4.2, where $_1X_2$ means the X force in element $\boxed{1}$ at node ② etc.

Thus for example the first equation is

$$_1X_1 = \frac{1}{0.866}(\tfrac{1}{4}d_{11}u_1 + 0.v_1 - \tfrac{1}{4}d_{11}u_2 + \sqrt{3}/4.d_{12}v_2 + 0u_3 - \sqrt{3}/4\,d_{12}\,v_3)$$

For plane stress $d_{11} = \dfrac{E}{1-v^2} = 1.0989$

$$\left.\begin{array}{c} \\ d_{12} = \dfrac{vE}{1-v^2} = 0.3297 \end{array}\right\} \quad \text{for } E = 1, v = 0.3$$

$$\therefore {}_1X_1 = 0.317u_1 + 0v_1 - 0.317u_2 + 0.165v_2 + 0.u_3 - 0.165v_3$$

The full six equations reduce to

$$
\begin{bmatrix} {}_1X_1 \\ {}_1Y_1 \\ {}_1X_2 \\ {}_1Y_2 \\ {}_1X_3 \\ {}_1Y_3 \end{bmatrix}
=
\begin{bmatrix}
0.317 & 0 & -0.317 & 0.165 & 0 & -0.165 \\
0 & 0.111 & 0.192 & -0.111 & -0.192 & 0 \\
-0.317 & 0.192 & 0.650 & -0.357 & -0.333 & 0.165 \\
0.165 & -0.111 & -0.357 & 1.063 & 0.192 & -0.952 \\
0 & -0.192 & -0.333 & 0.192 & 0.333 & 0 \\
-0.165 & 0 & 0.165 & -0.952 & 0 & 0.952
\end{bmatrix}
\begin{bmatrix} u_1 \\ v_1 \\ u_2 \\ v_2 \\ u_3 \\ v_3 \end{bmatrix}
$$

$$(4.15)$$

Element $\boxed{2}$: Consider the element to be numbered anticlockwise as ⑤,③,②. Thus in order to refer back to equations (4.14) say ①=⑤, ②=③ and ③=②.

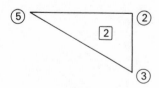

Thickness = 1
Area = 0.2165

Thus

$x_5 = x'_1 = 0$	$y_5 = y'_1 = 1$
$x_3 = x'_2 = \sqrt{3}/2$	$y_3 = y'_2 = \frac{1}{2}$
$x_2 = x'_3 = \sqrt{3}/2$	$y_2 = y'_3 = 1$

$$a_1 = x'_3 - x'_2 = 0 \qquad b_1 = y'_2 - y'_3 = -\tfrac{1}{2}$$

$$a_2 = x'_1 - x'_3 = -\sqrt{3}/2 \qquad b_2 = y'_3 - y'_1 = 0$$

$$a_3 = x'_2 - x'_1 = \sqrt{3}/2 \qquad b_3 = y'_1 - y'_2 = \tfrac{1}{2}$$

Proceeding in the same way as with element $\boxed{1}$ gives

$$
\begin{bmatrix} {}_2X_5 \\ {}_2Y_5 \\ {}_2X_3 \\ {}_2Y_3 \\ {}_2X_2 \\ {}_2Y_2 \end{bmatrix}
\begin{bmatrix}
0.317 & 0 & 0 & 0.165 & 0.317 & -0.165 \\
 & 0.111 & 0.192 & 0 & -0.192 & -0.111 \\
 & & 0.333 & 0 & -0.333 & -0.192 \\
 & & & 0.952 & -0.165 & -0.952 \\
 & & & & 0.650 & 0.357 \\
 & \text{Symmetrical} & & & & 1.063
\end{bmatrix}
\begin{bmatrix} u_5 \\ v_5 \\ u_3 \\ v_3 \\ u_2 \\ v_2 \end{bmatrix}
$$

$$(4.16)$$

Note that in element $\boxed{2}$, $u_5 = v_5 = 0$ and so columns one and two of the element stiffness matrix could be eliminated. In addition the first two equations are of no use initially, only after the solution of the unknown displacements.

Element $\boxed{3}$:

Consider the element to be numbered anticlockwise as ⑤, ④, ③ and as with element $\boxed{2}$ let ①≡⑤, ②≡④, ③≡③

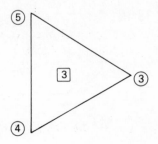

Thus

$x_5 = x'_1 = 0$	$y_5 = y'_1 = 1$
$x_4 = x'_2 = 0$	$y_4 = y'_2 = 0$
$x_3 = x'_3 = \sqrt{3}/2$	$y_3 = y'_3 = \tfrac{1}{2}$

$$a_1 = x'_3 - x'_2 = \sqrt{3}/2 \quad b_1 = y'_2 - y'_3 = -\tfrac{1}{2}$$

$$a_2 = x'_1 - x'_3 = -\sqrt{3}/2 \quad b_2 = y'_3 - y'_1 = -\tfrac{1}{2}$$

$$a_3 = x'_2 - x'_1 = 0 \quad b_3 = y'_1 - y'_2 = 1$$

Proceeding in the same way as with element 1 gives (4.17):

$$
\begin{bmatrix} {}_3X_5 \\ {}_3Y_5 \\ {}_3X_4 \\ {}_3Y_4 \\ {}_3X_3 \\ {}_3Y_3 \end{bmatrix}
=
\begin{bmatrix}
0.235 & -0.179 & -0.008 & -0.014 & -0.317 & 0.192 \\
 & 0.531 & 0.014 & -0.420 & 0.165 & -0.111 \\
 & & 0.325 & 0.179 & -0.317 & -0.192 \\
 & & & 0.531 & -0.165 & -0.111 \\
\text{Symmetrical} & & & & 0.634 & 0 \\
 & & & & & 0.222
\end{bmatrix}
\begin{bmatrix} u_5 \\ v_5 \\ u_4 \\ v_4 \\ u_3 \\ v_3 \end{bmatrix}
$$

In element ③ $u_4 = v_4 = u_5 = v_5 = 0$ and so columns one to four of the element stiffness matrix could be eliminated.

There are in this problem 6 unknown nodal displacements (degrees of freedom) and thus 6 equations are required. These equations come from applying equilibrium at the node of the unknown displacement and in the corresponding direction. Since the unknowns are u_1, v_1, u_2, v_2, u_3, v_3 equilibrium in both x- and y- directions at nodes ①, ② and ③ must be applied.

Equilibrium at node ①

Equilibrium of externally applied nodal force X_1 and element applied force ${}_1X_1$ give that ${}_1X_1 = X_1 = 0$ for this problem.

Thus from (4.15)

$$0.317u_1 + 0v_1 - 0.317u_2 + 0.165v_2 + 0.u_3 - 0.165v_3 = 0 \quad (4.18)$$

In the Y- direction ${}_1Y_1 = Y_1 = -1$ for this problem.

Thus from (4.15)

$$0 \, u_1 + 0.111v_1 + 0.192u_2 - 0.111v_2 - 0.192u_3 + 0v_3 = -1 \qquad (4.19)$$

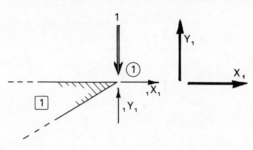

Equilibrium at node ②

In the X- direction

$${}_1X_2 + {}_2X_2 = X_2 = 0$$

Therefore, from (4.15) and (4.16)

$$-(0.317) \, u_1 + (0.192) \, v_1 + (0.650+0.650)u_2$$
$$+(-0.357+0.357)v_2 + (-0.333-0.333)u_3 + (0.165-0.165)v_3 = 0$$

or

$$-0.317u_1 + 0.192v_1 + 1.300u_2 + 0 \, v_2 - 0.666u_3 + 0v_3 = 0$$

$$(4.20)$$

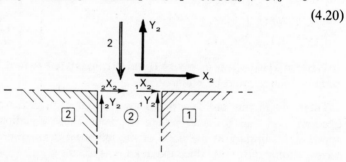

In the Y-direction

$${}_1Y_2 + {}_2Y_2 = Y_2 = -2$$

Thus from (4.15) and (4.16)

$$0.165u_1 - 0.111v_1 + 0.u_2 + 2.125v_2 + 0u_3 - 1.903v_3 = -2 \qquad (4.21)$$

Equilibrium at node ③

In the X-direction

$${}_1X_3 + {}_2X_3 + {}_3X_3 = X_3 = 0$$

And so from (4.15), (4.16) and (4.17)

$$0u_1 - 0.192v_1 - 0.666u_2 + 0v_2 + 1.301u_2 + 0v_3 = 0 \qquad (4.22)$$

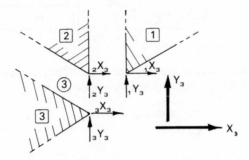

In the Y direction

$$_1Y_3 + _2Y_3 + _3Y_3 = Y_3 = 0$$

Therefore from (4.15), (4.16) and (4.17)

$$-0.165u_1 + 0v_1 + 0u_2 - 1.903v_2 + 0u_3 + 2.125v_3 = 0 \qquad (4.23)$$

Equations (4.18) to (4.23) are six linear algebraic simultaneous equations in the six unknowns u_1 to v_3: these can be put into matrix form

$$\begin{bmatrix} 0.317 & 0 & -0.317 & 0.165 & 0 & -0.165 \\ 0 & 0.111 & 0.192 & -0.111 & -0.192 & 0 \\ -0.317 & 0.192 & 1.300 & 0 & -0.666 & 0 \\ 0.165 & -0.111 & 0 & 2.125 & 0 & -1.903 \\ 0 & -0.192 & -0.666 & 0 & 1.301 & 0 \\ -0.165 & 0 & 0 & -1.903 & 0 & 2.125 \end{bmatrix} \begin{bmatrix} u_1 \\ v_1 \\ u_2 \\ v_2 \\ u_3 \\ v_3 \end{bmatrix} = \begin{bmatrix} 0 \\ -1 \\ 0 \\ -2 \\ 0 \\ 0 \end{bmatrix}$$

$$(4.24)$$

The solution to these equations yields

$$u_1 = 7.712 \qquad u_2 = 6.542 \qquad u_3 = -2.686$$
$$v_1 = -40.823 \qquad v_2 = -15.835 \qquad v_3 = -13.582 \left.\right\} (4.25)$$

In order to find the stresses in each element, the displacement vector for each element could be established $\{u'_1\, v'_1\, u'_2\, v'_2\, u'_3\, v'_3\}$ and the values put into equations (4.11) which result from the product $[\mathbf{D}]\ [\mathbf{B}]\ \{\mathbf{U}\}$. The results would give

	Element 1	Element 2	Element 3
σ_x	0	6.816	-3.408
σ_y	-4.505	-2.461	-1.022
τ_{xy}	-4	0.065	-6.032

(4.26)

Finally, in order to find the reaction forces at nodes ⑤ and ④, the equilibrium equations can be applied at these nodes as follows: ·

$$\text{At node ④} \quad {}_3X_4 = X_4$$
$$\qquad\qquad {}_3Y_4 = Y_4$$
$$\text{At node ⑤} \quad {}_2X_4 + {}_3X_5 = X_5$$
$$\qquad\qquad {}_2Y_5 + {}_3Y_5 = Y_5$$

(4.27)

The newly found values for u_2, v_2, u_3, v_3 from (4.25) and the boundary conditions $u_4 = v_4 = u_5 = v_5 = 0$, can be substituted into the 'unused' equations for ${}_3X_4$ etc. in equations (4.16) and (4.17).

This then gives $\quad -X_4 = 3.464 \quad X_5 = -3.464$
$$\qquad\qquad\qquad Y_4 = 1.951 \quad Y_5 = 1.049$$

(4.28)

From Fig. 4.6, these reactions can be seen to supply reactions which are in equilibrium with the applied forces.

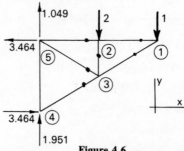

Figure 4.6

Notes. Several points should be noted about this very simple 3-element solution to the cantilever problem.

(a) The results in (4.25), (4.26) and (4.28) have been given no dimensions. The dimensions are in fact dependent on the dimensions of E, which defines force and length units, and thus the units of applied loading and node coordinates. The output is in the same units.

(b) The discretization (the dividing up into elements), is very crude and in fact leads to misleading results. Along the edge ④ ⑤, Fig. 4.6, a bending moment is being reacted and thus a distribution of bending stress σ_x from tensile at ⑤ through a neutral axis and to compressive at ④, would be expected. However, the element ③, being a constant strain (and thus stress) element, gives only one value of $\sigma_x = -3.408$ (from 4.26), which is clearly wrong. Stress values are usually ascribed to the element centroid. The solution to this dilemma is either to use higher-order elements, or use many more elements.

For example, in Fig. 4.7, a 32 element discretization (a) has 4 elements along the edge BC and when the stresses at the centroid of these elements are plotted (b), the smoothed curves start to show a more recognizable trend.

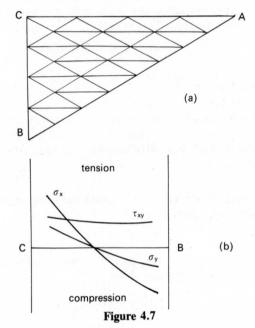

Figure 4.7

4.3 Implementation in program

4.3.1 *Description of program*

The program **FEP** is given in full in section 4.3.3 below. The data must be inserted into the program in DATA statements between lines 4000 and 10000. This form of data input means that the data is held within the program and can be stored on disc or tape with the program for future running and correction if required.

Stiffness matrices. The program deals with each element in turn starting with element number ①. Depending whether plane stress or plane strain conditions are required, PS is set to 1 or 2 respectively and thus the corresponding matrix

$[\mathbf{D}]$ established. Using the coordinates of the three nodes the $[\mathbf{B}]$ matrix is now filled up and the product calculated

$$[\mathbf{ST}] = [\mathbf{D}]\,[\mathbf{B}]$$

The array (matrix) $[\mathbf{ST}]$ for each element is stored for future retrieval and use. The second product leads to the element stiffness matrix being found from

$$[\mathbf{KL}] = t/2 \times [\mathbf{B}] \times [\mathbf{ST}]$$

The matrix $[\mathbf{KL}]$ is stored in $[\mathbf{KH}]$ for use at the end of the program.

However as with the program **PLFRAME**, the element stiffness matrix must be incorporated into the main structure stiffness matrix, which is similarly in condensed form. The restraints RE (I,1) and (I,2) are indicated as follows:

RE(I,1) = 1 indicates u at node I is zero
$-$ else = 0
RE(I,2) = 1 indicates v at I is zero
$-$ else = 0

The 6 values of RE for each element indicate which rows and columns have to be transferred into the structure stiffness matrix $[\mathbf{KS}]$, their locations in $[\mathbf{KS}]$ being indicated by the array MN, which is evaluated at the beginning of the program. The final size of the matrix $[\mathbf{KS}]$ is IN $\times$ IN.

Load vector. From the nodal force data at each node an uncondensed load vector is established in $\{\mathbf{P}\}$. Corresponding to the restraints RE this is progressively condensed and still held in $\{\mathbf{P}\}$ being finally of length IN.

Solution procedure. As with the two previous programs the same solution procedure is follows to solve the equations

$$[\mathbf{KS}]\,\{\mathbf{U}\} = \{\mathbf{P}\}$$

The solution of the unknown displacements is held in $\{\mathbf{P}\}$ and the vector of all nodal displacements is held in $\{\mathbf{PH}\}$, which is $\{\mathbf{P}\}$ with all the zero displacements added. This is a useful form from which to extract the displacement vectors for each element and also for finding the nodal forces.

Element stresses. Each element is dealt with in turn. From $\{\mathbf{PH}\}$ the 6 $\times$ 1, element displacement vector is extracted and established in $\{\mathbf{P}\}$ and the product of $\{\mathbf{P}\}$ and the retrieved matrix $[\mathbf{ST}]$ for the element produces the three stresses σ_x, σ_y and τ_{xy}

$$\{\sigma\} = [\mathbf{ST}].\,\{\mathbf{P}\}$$

For plane stress $\sigma_z = 0$ and for plane strain $\sigma_z = v(\sigma_y + \sigma_x)$. The program does not calculate σ_z. The strain can be worked out from the stresses from

$$\epsilon_x = \frac{1}{E} \left[\sigma_x - \nu \left(\sigma_y + \sigma_z \right) \right] \text{ etc}$$

Note: the stress values are usually assumed to be at the element centroid.

Nodal forces. As a good check on the solution and in order to find out the forces at the reaction points (the points of zero nodal displacements), the uncondensed stiffness matrix of the structure must first be established. This is done by retrieving for each element its 6 × 6 stiffness matrix, which was stored in $\left[\mathbf{KH} \right]$ and locating it in $\left[\mathbf{KS} \right]$, which is now the *uncondensed* structure stiffness matrix. The process is as if all values of RE = 0! The product of $\left[\mathbf{KS} \right]$ $\{ \mathbf{PH} \}$ now gives values of X- and Y- nodal forces at each node. Note the sum of all the X- forces should sum to zero, by equilibrium, as should the sum of the Y- forces. Also at loaded nodes the value of X- or Y- force should equal the data value input. At boundary nodes of zero displacement (RE(I,1) = 1 and/or RE(I,2) =1) the X- and/or Y-reaction forces can thus be found. All other nodal forces should be zero, subject to round-off errors etc.

4.3.2 *Data preparation*

A data preparation sheet is found in Appendix 4.1. The data is required in the following order, and, where more than one item of data is required for entry on any line, a comma is used as a separator.

Grid data:
(i)   Number of elements in the grid, NE
(ii)   Number of nodes in the grid, NN. This includes all boundary nodes.
(iii) Value of PS = 1 or 2 for plane stress or strain.

Note, as written, the program will solve for grids of up to 18 elements and 21 nodes with 42 degrees of freedom. To change these limits, EX and NX should be changed.

Nodal data: For each node, in any order.
(iv)   Node number – start off with first node as ① and there should be no gaps to NN
(v)   Coordinates x, y, of the node
(vi)   Restraints,   RE(I,1) = 1 for u_I = 0, else = 0
                     RE(I,2) = 1 for v_I = 0, else = 0
(vii) Applied nodal loads:
       X- direction force
       Y- direction force

Element data: For each element, in any order
(viii) Element number – start off with first element as ☐1 and there should be no gaps to NE

(ix)   Element node numbers – must be given anticlockwise

(x)   Element properties –   t – thickness
                              E – modulus of elasticity
                              NU – Poissons ratio

Notes (a)   The element numbers and the node numbers should start off at one
              and end with the last number, although the data may be input in
              any element order or node order.

     (b)   Units are as decided by the user but must be consistent. For
           example if E is in units of N/mm^2 then the thickness and the
           coordinates x and y are in mm, and the applied forces are in
           newtons. The output displacements will thus be in mm, the
           stresses in N/mm^2 and the nodal forces in newtons.

     (c)   It is recommended that element and nodal data be carefully
           checked when printed out at the beginning of the output, before
           allowing the program to proceed.

     (d)   Elements should have good aspect ratios, as near equilateral as
           possible.

4.3.3   Program *FEP*

Flow chart for **FEP**

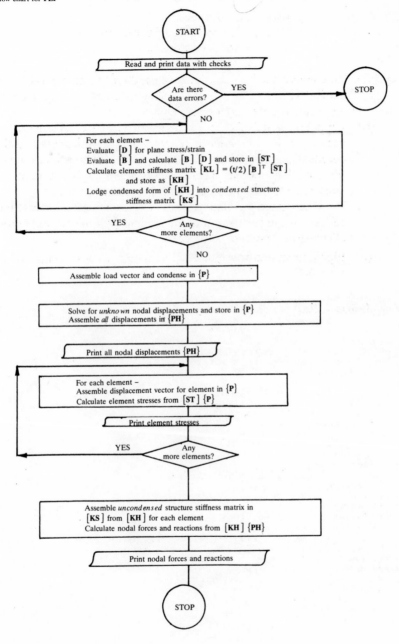

START

Read and print data with checks

Are there data errors? — YES → STOP

NO

For each element –
Evaluate $[D]$ for plane stress/strain
Evaluate $[B]$ and calculate $[B][D]$ and store in $[ST]$
Calculate element stiffness matrix $[KL] = (t/2)[B]^T[ST]$
   and store as $[KH]$
Lodge condensed form of $[KH]$ into *condensed* structure
   stiffness matrix $[KS]$

YES ← Any more elements?

NO

Assemble load vector and condense in $\{P\}$

Solve for *unknown* nodal displacements and store in $\{P\}$
Assemble *all* displacements in $\{PH\}$

Print all nodal displacements $\{PH\}$

For each element –
Assemble displacement vector for element in $\{P\}$
Calculate element stresses from $[ST]\{P\}$

Print element stresses

YES ← Any more elements?

Assemble *uncondensed* structure stiffness matrix in
$[KS]$ from $[KH]$ for each element
Calculate nodal forces and reactions from $[KH]\{PH\}$

Print nodal forces and reactions

STOP

```
100 REM                  *** PROGRAM  FEP ***
105 REM          AUTHOR:
110 REM               DAVID K. BROWN
112 REM     DEPARTMENT OF MECHANICAL ENGINEERING
115 REM             UNIVERSITY OF GLASGOW
120 REM                   SCOTLAND
125 REM
126 REM                 AUGUST  1983
127 REM
130 REM    **   PROGRAM WILL ANALYSE PLANE STRESS OR
132 REM         PLANE STRAIN PROBLEMS WITH UP TO
134 REM         18 ELEMENTS, 21 NODES AND 42 DEGREES
135 REM         OF FREEDOM USING CONSTANT STRAIN
136 REM         TRIANGLES  (CST).                    **
137 REM
138 REM    **   LOADING IS APPLIED THROUGH NODAL
139 REM         FORCES AND BOUNDARY DISPLACEMENTS
140 REM         ARE EITHER ZERO OR FREE.             **
145 REM
150 REM    **   INPUT TO THE PROGRAM IS THROUGH
152 REM         DATA STATEMENTS BETWEEN LINES
154 REM         4000   AND   10000.                 **
156 REM
158 REM    **   OUTPUT CONSISTS OF:
160 REM              NODAL DISPLACEMENTS U AND V,
165 REM              ELEMENT STRESSES AND
170 REM              NODAL FORCES.                   **
174 REM
176 REM
180 EX=18:NX=21 :REM  MAX # OF ELEMENTS AND MAX # OF NODES
190 NF=2*NX      :REM MAX # OF DEGREES OF FREEDOM
200 DIM X(NX),Y(NX),RE(NX,2),NJ(NX,3),T(EX),E(EX),NU(EX)
210 DIM P(NF),B(3,6),BT(6,3),D(3,3),ST(EX,3,6),PH(NF)
220 DIM KL(6,6),KS(NF,NF), KH(EX,6,6),MN(NX,2)
230 GOSUB 51000 :REM OPEN PRINTER CHANNEL
235 GOSUB 53000 :REM MACHINE SPECIFIC STRINGS
240 P$=" PROGRAM FEP":GOSUB50000
250 S1$="         ":REM 9 SPACES
260 S2$="           ":REM 11 SPACES
270 FORI=1TONF:FORJ=1TONF:KS(I,J)=0.:NEXTJ:NEXTI
280 REM  **   READ IN DATA STATEMENTS              **
290 READ NE,NN,PS
300 IFNE>EX THEN PRINT "TOO MANY ELEMENTS...>",EX
310 IF NE>EX THEN 2080
320 IFNN>NX THEN PRINT "TOO MANY NODES...>",NX
330 IF NN>NX THEN2080
340 P$="         ****** DATA INPUT ******"
345 GOSUB50000:IP=2:GOSUB50020
350 P$="NUMBER OF ELEMENTS = "+STR$(NE):GOSUB50000
355 P$="NUMBER OF NODES    = "+STR$(NN):GOSUB50000
360 P1$="PLANE STRESS CONDITIONS APPLY"
361 P2$="PLANE STRAIN CONDITIONS APPLY"
365 IF PS<>2 THEN PS=1:P$=P1$:GOSUB50000:IP=2:GOSUB50020
370 IF PS=2 THEN P$=P2$:GOSUB50000:IP=2:GOSUB50020
380 P$="*** NODAL DATA ***":GOSUB50000:IP=2:GOSUB50020
390 P$="NODE        COORDINATES           "
391 P$=P$+"RESTRAINTS      APPLIED FORCES"
392 GOSUB50000
```

```
400 P$=" NO            X          Y          U"
405 P$=P$+"         V          PX          PY"
406 GOSUB 50000
410 FORI=1TONN
420 READ N,X(I),Y(I),RE(I,1),RE(I,2),P(2*I-1),P(2*I)
425 P$=STR$(N)+"       "
430 XS=X(I):FW=11:NS=3:GOSUB20040:P$=P$+XS$
440 XS=Y(I):GOSUB20040:P$=P$+XS$+"      "
450 P$=P$+STR$(RE(I,1))+"        "+STR$(RE(I,2))+"      "
470 XS=P(2*I-1):GOSUB20040:P$=P$+XS$
480 XS=P(2*I):GOSUB20040:P$=P$+XS$
490 GOSUB50000:NEXT I
500 IP=3:GOSUB50020:P$="*** ELEMENT DATA ***"
505 GOSUB50000:IP=2:GOSUB50020
510 P$="ELEMENT    NODES    THICKNESS       ELASTIC    POISSON'S"
515 GOSUB50000
520 P$=" NO     1  2  3                  MODULUS     RATIO"
525 GOSUB 50000:GOSUB 50010
530 FORI=1TONE
540 READ N,NJ(I,1),NJ(I,2),NJ(I,3),T(I),E(I),NU(I)
545 P$="  "+STR$(N)+"      "+STR$(NJ(I,1))+" "
546 P$=P$+STR$(NJ(I,2))+" "+STR$(NJ(I,3))+" "
550 XS=T(I):GOSUB20000:P$=P$+XS$
560 XS=E(I):GOSUB10000:P$=P$+XS$
570 XS=NU(I):GOSUB20000:P$=P$+XS$
580 GOSUB50000
590 NEXTI
600 GOSUB 52000 :REM CLOSE PRINTER CHANNEL
610 PRINT "DO YOU WISH TO CORRECT THE DATA [Y/N]   ";N$;L4$;
615 INPUT AN$
620 IF LEFT$(AN$,1)="Y"THEN GOSUB51000:GOTO 2080
630 IF LEFT$(AN$,1)<>"N"THEN 610
640 PRINT "THE PROGRAM IS NOW RUNNING"
650 GOSUB 51000 :REM OPEN PRINTER CHANNEL
660 IP=4:GOSUB50020
670 P$="          ****** OUTPUT RESULTS ******"
675 GOSUB 50000
680 IN=0:FORI=1TONN: FORJ=1TO2
690 IF RE(I,J)=1 THEN 710
700 IN=IN+1: MN(I,J)=IN
710 NEXTJ
720 NEXTI
730 REM  ** \ CONDENSED STRUCTURE STIFFNESS MATRIX'
735 REM     / IS MK X MK IN SIZE.                    **
740 MK=IN
750 REM  **    SCAN THROUGH ALL ELEMENTS            **
760 FORIJK=1TONE
770 REM  **   DETERMINE CONSTITUTIVE MATRIX [D]
775 REM     (DEPENDING ON PLANE STRESS OR STRAIN) **
780 D(1,3)=0: D(2,3)=0:D(3,1)=0:D(3,2)=0
790 IF PS=2 THEN GOTO 850
800 U=E(IJK)/(1-NU(IJK)↑2)
810 D(1,1)=U: D(2,2)=U
820 D(2,1)=NU(IJK)*U: D(1,2)=NU(IJK)*U
830 D(3,3)=E(IJK)/2/(1+NU(IJK))
840 GOTO890
850 U=E(IJK)*(1-NU)/(1+NU(IJK))/(1-2*NU(IJK))
860 D(1,1)=U:D(2,2)=U: D(1,2)=NU(IJK)*U
870 D(2,1)=NU(IJK)*U: D(3,3)=E(IJK)/2/(1+NU(IJK))
880 REM  **   INITIALIZE MATRICES                    **
```

```
890 FORI=1TO6: FORJ=1TO6
900 KL(I,J)=0: NEXTJ
910 FORJ=1TO3: B(J,I)=0:BT(I,J)=0:ST(IJK,J,I)=0
920 NEXTJ: NEXTI
930 REM   ** \DETERMINE [B] MATRIX - NOT YET
935 REM        DIVIDED THROUGH BY (2*AREA)          **
940 N1=NJ(IJK,1):N2=NJ(IJK,2):N3=NJ(IJK,3)
950 X1=X(N1):X2=X(N2):X3=X(N3)
960 Y1=Y(N1): Y2=Y(N2): Y3=Y(N3)
970 AR=0.5*(Y1*X3-Y3*X1+Y3*X2-Y2*X3+Y2*X1-Y1*X2)
980 A1=X3-X2:A2=X1-X3:A3=X2-X1
990 B1=Y2-Y3:B2=Y3-Y1:B3=Y1-Y2
1000 B(1,1)=B1:B(1,3)=B2:B(1,5)=B3
1010 B(2,2)=A1:B(2,4)=A2:B(2,6)=A3
1020 B(3,1)=A1:B(3,3)=A2:B(3,5)=A3
1030 B(3,2)=B1:B(3,4)=B2:B(3,6)=B3
1040 REM   **   DETERMINE PRODUCT 0.5/A*[D][B] AND STORE
1050 REM        IN [ST] FOR FUTURE STRESS EVALUATION **
1060 FORI=1TO3:FORJ=1TO6:FORK=1TO3
1070 ST(IJK,I,J)=ST(IJK,I,J)+D(I,K)*0.5/AR*B(K,J)
1080 NEXTK:NEXTJ:NEXTI
1090 REM   **   DETERMINE ELEMENT STIFFNESS MATRIX [KL]
1095 REM        AND STORE AS [KH].                 **
1100 FORI=1TO6:FORJ=1TO6:FORK=1TO3
1110 KL(I,J)=KL(I,J)+T(IJK)/2*B(K,I)*ST(IJK,K,J)
1120 NEXTK:KH(IJK,I,J)=KL(I,J):NEXTJ:NEXTI
1130 REM   **   DEPENDING ON BOUNDARY CONDITIONS,
1135 REM        STORE STIFFNESS COEFFICIENTS INTO
1140 REM        STRUCTURE CONDENSED MATRIX [KS].    **
1150 FORI=1TO3:ND=NJ(IJK,I):IS=2*I-1
1160 FOR L=1 TO 2:IF RE(ND,L)=1 THEN 1240
1170 PK=MN(ND,L)
1180 FORJ=1TO3:JS=2*J-1:NC=NJ(IJK,J)
1190 FOR M=1 TO 2:IF RE(NC,M)=1 THEN 1220
1200 PL=MN(NC,M)
1210 KS(PK,PL)=KS(PK,PL)+KL(IS+L-1,JS+M-1)
1220 NEXTM
1230 NEXTJ
1240 NEXTL
1250 NEXTI
1260 NEXTIJK
1270 REM   **   ASSEMBLE LOAD VECTOR AND THEN
1275 REM        CONDENSE IN [P].                   **
1280 OT=0:FORI=1TONN:FORJ=1TO2
1290 IF RE(I,J)=0 THEN GOTO 1330
1300 FORM=2*I-(2-J)-OT TO 2*NN-OT-1
1310 P(M)=P(M+1)
1320 NEXTM: OT=OT+1
1330 NEXTJ
1340 NEXTI
1350 REM   **   SOLVE FOR UNKNOWN NODAL DISPLACEMENTS
1355 REM        AND STORE IN [P].                  **
1360 M=2*NN-OT:M1=M-1
1370 FORI=1TOM1: L=I+1
1380 FORJ=LTOM
1390 IF KS(J,I)=0 THEN GOTO 1440
1400 FORKK=LTOM
1410 KS(J,KK)=KS(J,KK)-KS(I,KK)*KS(J,I)/KS(I,I)
1420 NEXTKK
1430 P(J)=P(J)-P(I)*KS(J,I)/KS(I,I)
```

```
1440 NEXTJ
1450 NEXTI
1460 P(M)=P(M)/KS(M,M)
1470 FORI=1TOM1: KK=M-I: L=KK+1
1480 FORJ=L TO M
1490 P(KK)=P(KK)-P(J)*KS(KK,J)
1500 NEXTJ
1510 P(KK)=P(KK)/KS(KK,KK)
1520 NEXTI
1530 REM  **    ASSEMBLE ALL NODAL DISPLACEMENTS INTO [PH] **
1540 FORI=1TO 30: PH(I)=0
1550 NEXTI
1560 IN=0:FORI=1TONN:FORJ=1TO2
1570 IF RE(I,J)=1 THEN GOTO 1590
1580 IN=IN+1: PH(2*I-2+J)=P(IN)
1590 NEXTJ
1600 NEXTI
1610 IP=3:GOSUB50020
1620 P$="*** VECTOR OF ALL DISPLACEMENTS ***"
1625 GOSUB 50000:GOSUB 50010
1630 P$="NODE              DISPLACEMENTS"
1635 GOSUB 50000
1640 P$="  NO            U               V"
1641 GOSUB 50000:GOSUB 50010
1643 FW=15:NS=3
1645 FOR I = 1 TO NN:P$=STR$(I)+"      "
1650 XS=PH(2*I-1):GOSUB20040:P$=P$+XS$
1660 XS=PH(2*I):GOSUB20040:P$=P$+XS$
1670 GOSUB50000
1680 NEXTI
1690 REM  **  SCAN THROUGH ALL ELEMENTS AND
1693 REM        ESTABLISH ELEMENT DISPLACEMENT VECTOR
1697 REM        IN [P] AND MULTIPLY BY [ST] FROM
1700 REM        STORE TO GIVE ELEMENT STRESSES.    **
1710 GOSUB50010:P$="*** ELEMENT STRESSES ***"
1715 GOSUB 50000:GOSUB 50010
1720 P$="ELEMENT            S T R E S S E S"
1725 GOSUB 50000
1730 P$="  NO              SX        SY        TXY"
1735 GOSUB 50000
1740 FOR IJK=1TONE
1750 FORI=1TO3:NB=NJ(IJK,I)
1760 FORJ=1TO2
1770 P(2*I-2+J)=PH(2*NB-2+J)
1780 NEXTJ:NEXTI
1790 P$="": FORI=1TO3:SS(I)=0
1800 FORJ=1TO6:SS(I)=SS(I)+ST(IJK,I,J)*P(J):NEXTJ
1810 XS=SS(I):GOSUB10000:P$=P$+XS$
1820 NEXT I
1825 P$="  "+STR$(IJK)+"           "+P$:GOSUB50000
1830 NEXTIJK
1835 REM  **  BUILD UP UNCONDENSED STIFFNESS MATRIX
1840 REM        INTO [KS] AND MULTIPLY BY [PH] TO
1845 REM        FIND NODAL FORCES.              **
1850 IP=3:GOSUB50020
1860 FORI=1TO30: FORJ=1TO30
1870 KS(I,J)=0:NEXTJ:NEXTI
1880 FORIJK=1TONE
1890 FORI=1TO3:I1=2*NJ(IJK,I)-1:IS=2*I-1
1900 FORL=0TO1
```

E

```
1910 FORJ=1TO3:J1=2*NJ(IJK,J)-1:JS=2*J-1
1920 FORM=0TO1
1930 KS(I1+L,J1+M)=KS(I1+L,J1+M)+KH(IJK,IS+L,JS+M)
1940 NEXTM:NEXTJ:NEXTL:NEXTI:NEXTIJK
1950 FORI=1TO2*NN:P(I)=0
1960 FORJ=1TO2*NN
1970 P(I)=P(I)+KS(I,J)*PH(J)
1980 NEXTJ:NEXTI
1990 P$="*** NODAL FORCES ***"
1995 GOSUB 50000:GOSUB 50010
2000 P$="NODE                  NODAL   FORCES"
2005 GOSUB 50000
2010 P$=" NO                 PX              PY"
2015 GOSUB 50000:GOSUB 50010
2018 FW = 15: NS = 3
2020 FORI=1TONN:P$=STR$(I)+"              "
2030 XS=P(2*I-1):GOSUB20040:P$=P$+XS$
2040 XS=P(2*I):GOSUB20040:P$=P$+XS$
2050 GOSUB50000
2060 NEXTI
2070 GOSUB50010
2080 P$="      **** END OF RUN OF PROGRAM  F E P  ****"
2085 GOSUB 50000:IP=5:GOSUB 50020
2100 GOSUB 52000 :REM CLOSE PRINTER CHANNEL
2110 END
4000 REM *************************
4010 REM DATA STATEMENTS LOCATED BETWEEN
4020 REM LINES 4000 AND 10000
4030 REM *************************
10000 REM            FORMATTING AND INPUT/OUTPUT
10001 REM                  SUBROUTINES BY
10002 REM                  DAVID A. PIRIE
10003 REM    DEPARTMENT OF AERONAUTICS & FLUID MECHANICS
10004 REM              UNIVERSITY OF GLASGOW
10005 REM                   SCOTLAND
10006 REM                 AUGUST  1983
10010 REM
10015 REM   ** FORMAT NUMERICAL OUTPUT IN
10020 REM        SCIENTIFIC NOTATION              **
10035 FW=12:NS=4
10040 WE =1E-30
10045 KE=0:KE$="":BL$="          ":B0$="00000000"
10050 F5=FW-NS-5:N3=NS+3:Z$="0.":AX=ABS(XS)
10052 IF AX<WE THEN XS$=LEFT$(BL$,F5)+Z$+LEFT$(BL$,N3):GOTO10095
10055 IFABS(XS)<.010RABS(XS)>=1E9THEN10080
10060 IFABS(XS)<10RABS(XS)>=10THENGOSUB10175
10065 GOSUB10110
10070 GOTO10095
10080 XS$=STR$(XS):KE$=RIGHT$(XS$,3):KE=VAL(KE$)
10085 XS=VAL(LEFT$(XS$,LEN(XS$)-4))
10090 GOSUB10110
10095 RETURN
10110 REM FORM O/P$
10115 GOSUB10145
10120 IFABS(XS)>=10THENGOSUB10175
10125 GOSUB10200
10130 GOSUB10225
10135 RETURN
10145 REM ROUNDOFF MANTISSA
10155 XR=5:FORI5=1TONS:XR=XR/10:NEXTI5.
```

```
10160 XS=XS+XR*SGN(XS)
10165 RETURN
10175 REM NORMALISE MANTISSA
10180 IF ABS(XS)<1THENXS=XS*10:KE=KE-1:GOSUB10180
10185 IF ABS(XS)>=10THENXS=XS/10:KE=KE+1:GOSUB10185
10190 RETURN
10200 REM FORM EXPONENT$
10205 S$="+":IFKE<0THENS$="-"
10210 KE$=S$+RIGHT$("0"+MID$(STR$(KE),2),2)
10215 RETURN
10225 REM FORM (MANT+EXP)$
10230 X1$=LEFT$(STR$(XS),NS+2)
10235 XS$=X1$+LEFT$(B0$,NS+2-LEN(X1$))
10240 IFXS=INT(XS)THEN XS$=X1$+"."+LEFT$(B0$,NS-1)
10245 XS$=LEFT$(BL$,FW-NS-6)+XS$+"E"+KE$
10250 RETURN
20000 REM  **  FORMAT NUMERICAL OUTPUT            **
20020 FW=12:NS=3
20040 BL$="                    "
20050 XS$=STR$(XS):XE$="    ":IFLEN(XS$)<4THENXS$=XS$+"    "
20060 IFABS(XS)>=10↑(8-NS)THENXX=XS:GOTO20080
20070 XX=XS+.5*SGN(XS)/10↑NS
20080 IFMID$(XS$,LEN(XS$)-3,1)="E"THENGOSUB20180
20090 XX$=STR$(XX)
20100 FORJ5=1TOLEN(XX$)
20110 IFMID$(XX$,J5,1)="."THENDP=J5:GOTO20130
20120 NEXTJ5:DP=LEN(XX$)+1:XX$=XX$+".0000000"
20130 XS$=LEFT$(XX$,DP+NS)+XE$
20140 LX=LEN(XS$):IFLX>FWTHENXS$=LEFT$(XS$,FW):GOTO20160
20150 XS$=LEFT$(BL$,FW-LX)+XS$
20160 RETURN
20180 XE$=RIGHT$(XS$,4):XR=VAL(RIGHT$(XS$,2))
20190 XX=VAL(LEFT$(XS$,LEN(XS$)-4))+.5*SGN(XS)/10↑NS
20200 RETURN
50000 REM ** THE FOLLOWING STATEMENTS
50001 REM    MUST BE TAILORED TO THE
50002 REM     PARTICULAR MACHINE IN USE        **
50005 PRINT#5,P$:RETURN :REM ** PRINTLINE ON PRINTER **
50010 PRINT#5:RETURN    :REM ** 1 LINEFEED ON PRINTER **
50015 REM
50020 FOR KP = 1 TO IP :REM IP
50021 PRINT#5            :REM    LINEFEEDS
50022 NEXT KP            :REM    ON
50023 RETURN            :REM    PRINTER *
50025 REM
51000 OPEN 5,4:RETURN   :REM ** OPEN CHANNEL TO PRINTER  **
52000 CLOSE 5:RETURN    :REM ** CLOSE CHANNEL TO PRINTER **
53000 REM ** THE FOLLOWING Y$,N$,L4$,AK$
53001 REM     ARE USED WITH 'INPUT' STATEMENTS -
53002 REM     SET THEM ALL EQUAL TO "" IF
53003 REM     L4$ NOT POSSIBLE ON YOUR MACHINE    **
53010 Y$="Y ":N$="N ":AK$="* "
53020 L4$="▮▮▮▮":RETURN :REM ** L4$ = 4 'CURSOR-LEFTS'  **
```

4.3.4 *Example of solution to continuum problem using **FEP***

(a) *The structure.* Symmetrical quarter of a thick plate with a circular hole in uniaxial tension.

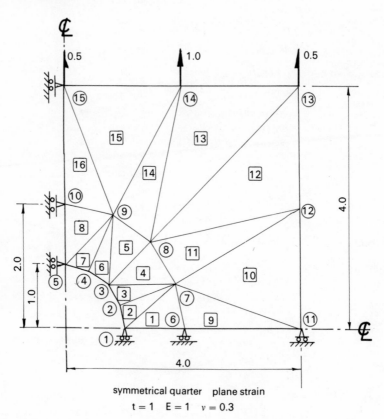

symmetrical quarter   plane strain

t = 1   E = 1 ν = 0.3

(b) *The data.*

```
4000 DATA16,15,2
4010 DATA1,1,0,0,1,0,0
4020 DATA2,.976,.216,0,0,0,0
4030 DATA3,.707,.707,0,0,0,0
4040 DATA4,.216,.976,0,0,0,0
4050 DATA5,0,1,1,0,0,0
4060 DATA6,2.0,0,0,1,0,0
4070 DATA7,1.848,.765,0,0,0,0
4080 DATA8,1.414,1.414,0,0,0,0
4090 DATA9,.765,1.848,0,0,0,0
4100 DATA10,0,2.0,1,0,0,0
4110 DATA11,4,0,0,1,0,0
4120 DATA12,4,2,0,0,0,0
4130 DATA13,2.828,2.828,0,0,0,.5
4140 DATA14,2,4,0,0,0,1
4150 DATA15,0,4,1,0,0,.5
```

```
4160 DATA1,1,6,7,1,1,.3
4170 DATA2,1,7,2,1,1,.3
4180 DATA3,2,7,3,1,1,.3
4190 DATA4,3,7,8,1,1,.3
4200 DATA5,3,8,9,1,1,.3
4210 DATA6,3,9,4,1,1,.3
4220 DATA7,4,9,5,1,1,.3
4230 DATA8,5,9,8,1,1,.3
4240 DATA9,6,11,7,1,1,.3
4250 DATA10,7,11,12,1,1,.3
4260 DATA11,7,12,8,1,1,.3
4270 DATA12,8,12,13,1,1,.3
4280 DATA13,8,13,14,1,1,.3
4290 DATA14,8,14,9,1,1,.3
4300 DATA15,9,14,15,1,1,.3
4310 DATA16,9,15,10,1,1,.3
```

(c) *The solution.*

```
****** OUTPUT RESULTS ******


*** VECTOR OF ALL DISPLACEMENTS ***

NODE          DISPLACEMENTS
 NO         U              V

  1      -.423         0.000
  2      -.366          .196
  3      -.256          .604
  4      -.091         1.208
  5      0.000         1.422
  6      -.483         0.000
  7      -.447          .287
  8      -.235          .655
  9      -.074          .892
 10      0.000         1.179
 11      -.652         0.000
 12      -.486          .336
 13      -.282          .868
 14      -.099         1.573
 15      0.000         1.611

*** ELEMENT STRESSES ***

ELEMENT            S  T  R  E  S  S  E  S
  NO         SX            SY           TXY
   1     1.016E-01     6.866E-01     1.339E-02
   2     2.517E-02     1.510E+00    -7.558E-02
   3     4.651E-02     1.169E+00    -7.069E-02
   4    -1.293E-01     6.071E-01    -3.403E-02
   5    -1.111E-01     4.249E-01    -8.804E-03
   6    -2.906E-01     4.487E-01    -3.425E-01
   7    -6.107E-01     2.462E-01    -2.694E-01
   8    -4.743E-01    -4.664E-01    -1.553E-01
   9     5.432E-02     6.725E-01     1.152E-02
  10    -2.873E-02     2.857E-01     3.631E-03
  11    -4.260E-02     7.242E-01     3.491E-03
  12    -1.802E-02     6.142E-01    -4.652E-02
  13     2.618E-02     7.339E-01    -7.173E-02
  14    -1.355E-01     6.243E-01    -6.109E-03
  15     9.378E-02     6.014E-01    -7.715E-04
  16    -6.238E-02     3.596E-01    -1.278E-01



*** NODAL FORCES ***

NODE            NODAL   FORCES
 NO         PX             PY

  1     1.457E-10      -.736
  2    -2.976E-11      2.501E-10
  3    -3.643E-10      0.000
```

```
PROGRAM FEP
        ****** DATA INPUT ******


NUMBER OF ELEMENTS =  16
NUMBER OF NODES    =  15
PLANE STRAIN CONDITIONS APPLY


*** NODAL DATA ***

NODE      COORDINATES        RESTRAINTS      APPLIED FORCES
 NO     X         Y          U     V         PX        PY
  1   1.000     0.000        0     1        0.000     0.000
  2    .976      .216        0     0        0.000     0.000
  3    .707      .707        0     0        0.000     0.000
  4    .216      .976        0     0        0.000     0.000
  5   0.000     1.000        1     0        0.000     0.000
  6   2.000     0.000        0     1        0.000     0.000
  7   1.848      .765        0     0        0.000     0.000
  8   1.414     1.414        0     0        0.000     0.000
  9    .765     1.848        0     0        0.000     0.000
 10   0.000     2.000        1     0        0.000     0.000
 11   4.000     0.000        0     1        0.000     0.000
 12   4.000     2.000        0     0        0.000     0.000
 13   2.828     2.828        0     0        0.000      .500
 14   2.000     4.000        0     0        0.000     1.000
 15   0.000     4.000        1     0        0.000      .500


*** ELEMENT DATA ***


ELEMENT    NODES    THICKNESS     ELASTIC    POISSON'S
  NO     1   2   3                MODULUS      RATIO

   1     1   6   7    1.000      1.000E+00     .300
   2     1   7   2    1.000      1.000E+00     .300
   3     2   7   3    1.000      1.000E+00     .300
   4     3   7   8    1.000      1.000E+00     .300
   5     3   8   9    1.000      1.000E+00     .300
   6     3   9   4    1.000      1.000E+00     .300
   7     4   9   5    1.000      1.000E+00     .300
   8     5   9   8    1.000      1.000E+00     .300
   9     6  11   7    1.000      1.000E+00     .300
  10     7  11  12    1.000      1.000E+00     .300
  11     7  12   8    1.000      1.000E+00     .300
  12     8  12  13    1.000      1.000E+00     .300
  13     8  13  14    1.000      1.000E+00     .300
  14     8  14   9    1.000      1.000E+00     .300
  15     9  14  15    1.000      1.000E+00     .300
  16     9  15  10    1.000      1.000E+00     .300


   4              3.235E-10      2.465E-10
   5               .039         1.113E-09
   6             -2.628E-10     -1.014
   7              0.000         1.009E-09
   8             -6.555E-10     -8.822E-10
   9              7.140E-10      4.529E-10
  10               .116        -3.311E-10
  11              2.673E-10      -.250
  12              1.305E-10     0.000
  13              2.392E-10      .500
  14              1.858E-10     1.000
  15              -.155          .500

      ****   END OF RUN OF PROGRAM  F E P   ****
```

Appendix 4.1 FEP (Computer program for continuum plane stress or plane strain analysis): program summary and data sheet

1. Introduction

From data read in regarding plane stress/strain, nodal coordinates and element nodes, the element stiffness matrix is calculated from

$$[\mathbf{K}] = \text{t.A} [\mathbf{B}]^T [\mathbf{D}] [\mathbf{B}]$$

Depending on zero displacement boundary conditions, the stiffness coefficients are transferred into the condensed structure stiffness matrix. The condensed load vector $\{\mathbf{P}\}$ is established and so the equations are solved to give all the unknown nodal displacements $\{\mathbf{U}\}$.

The stresses in each element are calculated from

$$\{\sigma\} = [\mathbf{D}] [\mathbf{B}] \{\mathbf{U}\}$$

Finally by assembling the uncondensed stiffness matrix and multiplying it by the full nodal displacement vector, all the nodal forces are found, thus determining the reaction forces (at the zero displacement nodes).

Preparation of input data for this program should be accomplished in the following sequence:

(1) Sketch the structure and discretize the continuum into a satisfactory number of elements, concentrating elements at areas of high stress gradients and trying to keep the triangles as equilateral as possible.
(2) Number the nodes and elements
(3) Establish a reference coordinate system and determine node coordinates
(4) Remember the element nodes should be fed into the program in an *anticlockwise* sense
   – as shown

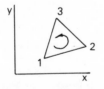

(5) Define the different load cases to be considered.
(6) Fill out datasheet.

2. Datasheet for **FEP**

Note: units must be consistent

Structure data. Number of elements (NE $\leq$ 18), Number of nodes (NN $\leq$ 21), plane stress or strain (1 or 2)

	,	,	

Nodal data. Node number, its coordinates, its restraints (= 1 for zero displacement, else = 0), and applied loads for each node starting with ① and ending with NN

Node No	Coordinates		Restraints on		Applied nodal loads	
	X	Y	u	v	P_x	P_y
,	,	,	,	,	,	
,	,	,	,	,	,	
,	,	,	,	,	,	
,	,	,	,	,	,	
,	,	,	,	,	,	
,	,	,	,	,	,	
,	,	,	,	,	,	
,	,	,	,	,	,	
,	,	,	,	,	,	
,	,	,	,	,	,	

Element data. Element number, its defining nodes, thickness elastic modulus and Poisson's ratio for each element starting with ⬚1 and ending with NE

Element No.	Node Numbers				Thickness T	Material Properties	
						E	NU
,	,	,	,	,	,	,	
,	,	,	,	,	,	,	
,	,	,	,	,	,	,	
,	,	,	,	,	,	,	
,	,	,	,	,	,	,	
,	,	,	,	,	,	,	
,	,	,	,	,	,	,	
,	,	,	,	,	,	,	

Notes: (a) This is merely a sample blank data sheet for up to 10 nodes and 8 elements.

(b) The data should be typed into the program in DATA statements between line numbers 4000 and 10000.

5 FEPB: Bending of thin flat plates

5.0 Introduction

In dealing with the computer analysis of thin flat plates a similar approach is taken here to that for the previous program **FEP**. Now, however, a rectangle is taken, so the element has four nodes, and at each node there exists three unknowns – a z-displacement w and two rotations or slopes about the x- and y-axis θx, θy. As before a function describing the variation of w across the element is assumed, in this case a fourth order polynomial in the local coordinates x and y, with the origin of the local axis system being located at the centre of element. A restriction exists on the formulation presented here: the sides of the element must be parallel to the structure (whole plate) axes. It is of note here that the stiffness matrix for the element is built up through working in the element's own local coordinate system and not as in **FEP**, where the global or structural coordinate system was used.

With higher order elements, analysis is always done in terms of the element's own or natural coordinates, whether triangles or quadrilaterals. Also in higher order elements, the solving of a large number of constants in the assumed polynomial function becomes excessive and an approach using interpolation functions is employed. However, such approaches lead to greater complications, which are beyond the scope of the current programs, and reference can be made to the texts suggested in Chapter 1.

Thus unlike **PJFRAME** and **PLFRAME,** this program has an approximation included from the start and so the greater the number of elements used, the greater will be the accuracy of the solution. To speed up **FEPB**, the principle of typing elements is used, in which elements with identical size and orientation are given the same type number because they have an identical stiffness matrix. Since, in **FEPB,** the derivation of stiffness matrices is time-consuming, only the first of any group of elements of the same type need have their stiffness matrix derived and others of the same type merely access and use these stored matrices.

One final word on this element. It has one further restriction, beyond its obvious rectangular shape – it does not take into account deformations due to shear forces along the edges. Although this is no real disadvantage for thin plates, the thicker they become the more important becomes the shear effect.

The usual assumptions of thin flat plate theory are used, and the basic relations can be found in any good mechanics book.

5.1 Development of equations

5.1.1 *Finite element method — rectangular element*

Consider a thin rectangular plate element ijkl of uniform thickness t with a *local coordinate* system 0xyz such that 0x and 0y are parallel to the sides of the element and positive z – direction is downwards. The origin is taken at the centre of the element at mid depth as shown in Fig. 5.1.

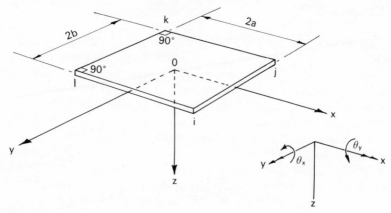

Figure 5.1

positive rotation about x- and y-axes

Assume a displacement function.[*]

$$w = \alpha_1 + \alpha_2 x + \alpha_3 y + \alpha_4 x^2 + \alpha_5 xy + \alpha_6 y^2 + \alpha_7 x^3 + \alpha_8 x^2 y$$

$$+ \alpha_9 xy^2 + \alpha_{10} y^3 + \alpha_{11} x^3 y + \alpha_{12} xy^3 \tag{5.1}$$

where w is the displacement of the plate in the positive z – direction.

Note that in thin-plate theory the in-plane or membrane forces are ignored and so all u- and v- displacements are zero and the z- displacement is due solely to bending.

The 'displacements' or degrees of freedom at each node i are w_i, the deflection, and θ_{xi} and θ_{yi}, the rotations.

$$\text{Thus let } \{\Delta_i\} = \begin{bmatrix} w_i \\ \theta_{xi} \\ \theta_{yi} \end{bmatrix}$$

$$\tag{5.2}$$

---

[*] Zienkiewicz, O.C. and Cheung, Y.K. The Finite element method of analysis of elastic isotropic and orthotropic slabs, *Proc. Inst. Civ. Engrs,* **28,** 471-488 (1964).

However, from plate theory $\theta_x = -\partial w/\partial x$, $\theta_y = +\partial w/\partial y$, and so for the element ijkl, the displacement vector is

$$
\{\Delta\} =
\begin{bmatrix} \Delta_i \\ \Delta_j \\ \Delta_k \\ \Delta_l \end{bmatrix}
=
\begin{bmatrix}
w_i \\ \theta_{xi} \\ \theta_{yi} \\ w_j \\ \theta_{xj} \\ \theta_{yj} \\ w_k \\ \theta_{xk} \\ \theta_{yk} \\ w_l \\ \theta_{xl} \\ \theta_{yl}
\end{bmatrix}
=
\begin{bmatrix}
w_i \\ -(\partial w/\partial x)_i \\ (\partial w/\partial y)_i \\ w_j \\ -(\partial w/\partial x)_i \\ (\partial w/\partial y)_i \\ w_k \\ -(\partial w/\partial x)_k \\ (\partial w/\partial y)_k \\ w_l \\ -(\partial w/\partial x)_l \\ (\partial w/\partial y)_l
\end{bmatrix}
\tag{5.3}
$$

Using the assumed displacement function (5.1) in (5.3) gives $\{\Delta\} = [A]\{\alpha\}$ or, written out fully

w_i	1	x_i	y_i	x^2_i	x_iy_i	y^2_i	x^3_i	$x^2_iy_i$	$x_iy^2_i$	y^3_i	$x^3_iy_i$	$x_iy^3_i$
θ_{xi}	0	-1	0	$-2x_i$	$3y_i$	0	$-3x^2_i$	$-2x_iy_i$	$-y^2_i$	0	$-3x^2_iy_i$	$-y^3_i$
θ_{yi}	0	0	1	0	x_i	$2y_i$	0	x^2_i	$2x_iy_i$	$3y^2_i$	x^3_i	$3x_iy^2_i$

with the remaining partitioned rows:

- w_j, θ_{xj}, θ_{yj} : As above with x_j, y_j
- w_k, θ_{xk}, θ_{yk} : As above with x_k, y_k
- w_l, θ_{xl}, θ_{yl} : As above with x_l, y_l

multiplied by the vector $\{\alpha_1, \alpha_2, \alpha_3, \alpha_4, \alpha_5, \alpha_6, \alpha_7, \alpha_8, \alpha_9, \alpha_{10}, \alpha_{11}, \alpha_{12}\}^T$

$$\tag{5.4}$$

The vector of unknown coefficients $\{\alpha\}$ can be found from the inverse of $[\mathbf{A}]$

$$\{\alpha\} = [\mathbf{A}]^{-1}\{\Delta\} \tag{5.5}$$

(Note that this procedure was done in (4.4) to (4.6) merely by solving the equations.) The vector of applied nodal forces is given by

$$\{\mathbf{P}\} = \begin{bmatrix} P_i \\ P_j \\ P_k \\ P_l \end{bmatrix} = \begin{bmatrix} P_{zi} \\ T_{xi} \\ T_{yi} \\ \vdots \\ \vdots \\ \vdots \\ P_{xl} \\ T_{xl} \\ T_{yl} \end{bmatrix} \tag{5.6}$$

where e.g. P_{zi} is force in z-direction at node i

T_{xi} is moment in xz plane about y-axis at node i

T_{yi} is moment in yz plane about x-axis at node i

The applied moments T act at nodal points and may be considered either as twisting or bending moments.

The sign convention for internal bending and twisting moments within element ijkl is shown in Fig.5.2. where, on the element dy × dx

M_x =   Bending moment/unit width on a positive x-face tending to produce a positive displacement w in the plate

M_y =   Bending moment/unit width on a positive y-face tending to produce a positive displacement w in the plate

M_{xy} =   Twisting moment/unit width on a positive x-face in the yz plane tending to product positive rotation about the x axis

M_{yx} =   Twisting moment/unit width on a positive y-face in the xz plane tending to produce positive rotation about the y-axis

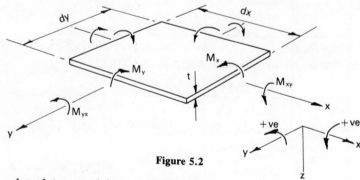

Figure 5.2

From plate theory, applying equilibrium gives

$$M_x = \int_{-t/2}^{t/2} \sigma_x z\,dz = -\frac{Et^3}{12(1-v^2)} \left\{ \frac{\partial^2 w}{\partial x^2} + v\,\frac{\partial^2 w}{\partial y^2} \right\}$$

$$M_y = \int_{-t/2}^{+t/2} \sigma_y z\,dz = -\frac{Et^3}{12(1-v^2)} \left\{ v\,\frac{\partial^2 w}{\partial x^2} + \frac{\partial^2 w}{\partial y^2} \right\}$$

$$M_{xy} = \int_{-t/2}^{+t/2} \tau_{xy} z\,dz = +\frac{Et^3}{12(1-v^2)} (1-v)\,\frac{\partial^2 w}{\partial x\,\partial y}$$

$$M_{yx} = \int_{-t/2}^{+t/2} \tau_{yx} z\,dz = -\frac{Et^3}{12(1-v^2)} (1-v)\,\frac{\partial^2 w}{\partial x\,\partial y} = -M_{xy}$$

Thus

$$
\begin{bmatrix} M_x \\ M_y \\ M_{yx} \end{bmatrix}
= \frac{Et^3}{12(1-v^2)}
\begin{bmatrix} +1 & +v & 0 \\ +v & +1 & 0 \\ 0 & 0 & \frac{1-v}{2} \end{bmatrix}
\begin{bmatrix} -\dfrac{\partial^2 w}{\partial x^2} \\ -\dfrac{\partial^2 w}{\partial y^2} \\ 2\dfrac{\partial^2 w}{\partial x\,\partial y} \end{bmatrix}
$$

$$(5.7)$$

$$or\ \{M\} = [D]\,\{C\}$$

where $\{C\}$ is the curvature vector for element $dx \times dy$. Since M_x, M_y, M_{xy} are moment/unit length, the moments acting on $dxdy$ are

$$M_x dy,\ M_y dx,\ M_{xy} dy,\ M_{yx} dx$$

Thus the strain energy $U = \frac{1}{2}\left\{ -M_x dy\ (\frac{\partial^2 w}{\partial x^2})\,dx - M_y\,dx\ (\frac{\partial^2 w}{\partial y^2})\,dy \right.$

$$\left. + 2M_{xy}\ (\frac{\partial^2 w}{\partial x\,\partial y})\,dxdy \right\}$$

or   for elemental area dxdy, $U = \frac{1}{2}\{C\}^T\{M\}\,dxdy$ (5.8)

However from (5.1), the curvatures can be found such that

$$
\{C\} = \begin{bmatrix} -\dfrac{\partial^2 w}{\partial x^2} \\[2mm] -\dfrac{\partial^2 w}{\partial y^2} \\[2mm] 2\dfrac{\partial^2 w}{\partial x \partial y} \end{bmatrix}
\begin{bmatrix}
0 & 0 & 0 & -2 & 0 & 0 & -6x & -2y & 0 & 0 & -6xy & 0 \\
0 & 0 & 0 & 0 & 0 & -2 & 0 & 0 & -2x-6y & 0 & -6xy \\
0 & 0 & 0 & 0 & 2 & 0 & 0 & 4x & 4y & 0 & 6x^2 & 6y^2
\end{bmatrix}
\begin{bmatrix} \alpha_1 \\ \alpha_2 \\ \alpha_3 \\ \vdots \\ \alpha_{12} \end{bmatrix}
$$

(5.9)

$$\text{or } \{C\} = [B]\{\alpha\}$$

But from (5.4) inversion leads to (5.5), viz. $\{\alpha\} = [A]^{-1}\{\Delta\}$

$$\text{and so } \{C\} = [B][A]^{-1}\{\Delta\}$$ (5.10)

and from (5.7) $\{M\} = [D]\{C\} = [D][B][A]^{-1}\{\Delta\}$ (5.11)

In order to relate $\{P\}$ (5.6) to $\{\Delta\}$, (5.3) through the stiffness matrix, use is made of the theorem of virtual work. The element stiffness matrix $[K]$ comes from

$$\{P\} = [K]\{\Delta\}$$ (5.12)

If the element ijkl is subjected to nodal displacements $\{\Delta\}$, there are required nodal forces $\{P\}$ as given by (5.12) and internal moment $\{M\}$ as given by (5.11). Now when virtual displacements $\{\Delta'\}$ are imposed these form a compatible deformation set with virtual internal curvatures such that

$$\{C'\} = [B][A]^{-1}\{\Delta'\}$$ (5.13)

Virtual work equation:

Work done by applied loads $= \frac{1}{2}\{\Delta'\}^T\{P\} = \frac{1}{2}\overbrace{\int_{-a}^{+a}\int_{-b}^{+b}\underbrace{\{C'\}^T\{M\}dxdy}_{\text{deformation set}}}^{\text{equilibrium set}}$

= internal strain energy

Using (5.11), (5.12) and (5.13) leads to

$$\{\Delta'\}^T [K] \{\Delta\} = \int_{-a}^{a} \int_{-b}^{b} \left\{ \{\Delta'\}^T [A]^{-T} [B]^T \right\} \left\{ [D][B][A]^{-1} \{\Delta\} \right\} dxdy$$

$$(\text{Note: } [A]^{-T} \equiv [A^{-1}]^T)$$

$$= \{\Delta'\}^T [A]^{-T} \left\{ \int_{-a}^{a} \int_{-b}^{b} [B]^T [D] [B] \, dxdy \right\} [A]^{-1} \{\Delta\}$$

Thus
$$[K] = [A]^{-T} \left\{ \int_{-a}^{a} \int_{-b}^{b} [B]^T [D] [B] \, dxdy \right\} [A]^{-1} \tag{5.14}$$

is the element stiffness matrix.

Given the nodal coordinates of the element, the matrix $[A]$, (5.4) can be evaluated and a standard procedure will produce the inverse of $[A]$, $[A]^{-1}$. The matrix $\int_{-a}^{a} \int_{-b}^{b} [B]^T [D] [B] \, dxdy$ can be found as shown below in 5.1.2 and is a 12×12. Thus the triple product of the three 12×12 matrices can be produced to give the 12×12 stiffness matrix of each element.

5.1.2   Determination of triple product

From (5.9) and (5.7)

$$[B]^T [D] = \frac{Et^3}{12(1-v^2)}
\begin{bmatrix}
0 & 0 & 0 \\
0 & 0 & 0 \\
0 & 0 & 0 \\
-2 & 0 & 0 \\
0 & 0 & 2 \\
0 & -2 & 0 \\
-6x & 0 & 0 \\
-2y & 0 & 4x \\
0 & -2x & 4y \\
0 & -6y & 0 \\
-6xy & 0 & 6x^2 \\
0 & -6xy & 6y^2
\end{bmatrix}
\begin{bmatrix}
1 & v & 0 \\
v & 1 & 0 \\
0 & 0 & \frac{1-v}{2}
\end{bmatrix}
= \frac{Et^3}{12(1-v^2)}
\begin{bmatrix}
0 & 0 & 0 \\
0 & 0 & 0 \\
0 & 0 & 0 \\
-2 & -2v & 0 \\
0 & 0 & 1-v \\
-2v & -2 & 0 \\
-6x & -6xv & 0 \\
-2y & -2yv & 2x(1-v) \\
-2xv & -2x & 2y(1-v) \\
-6yv & -6y & 0 \\
-6xv & -6xyv & 3x^2(1-v) \\
-6xyv & -6xy & 3y^2(1-v)
\end{bmatrix}$$

$$([\mathbf{B}]^{\mathrm{T}}[\mathbf{D}]).[\mathbf{B}] = ([\mathbf{B}]^{\mathrm{T}}[\mathbf{D}])\begin{bmatrix} 0 & 0 & 0 & -2 & 0 & 0 & -6x & -2y & 0 & 0 & -6xy & 0 \\ 0 & 0 & 0 & 0 & 0 & -2 & 0 & 0 & -2x & -6y & 0 & -6xy \\ 0 & 0 & 0 & 0 & 2 & 0 & 0 & 4x & 4y & 0 & 6x^2 & 6y^2 \end{bmatrix}$$

Thus (5.14) follows (see Table 5.1) and integration leads to (5.15) (see Table 5.2).

5.1.3 *Assemblage and solution*

In any plate problem consisting of several elements, the stiffness matrices of each element can thus be found from (5.14), remembering that the x's and y's in $[\mathbf{A}]$ are in local coordinates, relative to a central origin. It can thus be seen that identically-sized elements will have identical stiffness matrices, assuming that each has the same thickness and elastic constants E, as required in (5.15).

In order to 'pull' the elements together, as in the three previous programs, nodal equilibrium is used and this will lead finally to the required number of linear algebraic simultaneous equations corresponding to the number of unknown w's and θ's. The coefficients of the w's and θ's form the structure stiffness matrix, which, with the applied load vector $\{\mathbf{P}\}$, is solved to give the vector of the unknown displacements $\{\mathbf{\Delta}\}$.

In order to find the internal bending and twisting moments within any element, $\{\mathbf{\Delta}\}$ is multiplied into (5.11):

$$\{\mathbf{M}\} = [\mathbf{D}][\mathbf{B}][\mathbf{A}]^{-1}\{\mathbf{\Delta}\}$$

where $\{\mathbf{\Delta}\}$ is the displacement vector for this particular element. Matrices $[\mathbf{D}]$ and $[\mathbf{A}]^{-1}$ are fixed and known for this element but the matrix $[\mathbf{B}]$, (5.9), is written in general x, y coordinates where $-a < x < +a, -b < y < +b$. Thus the moments M_x, M_y, M_{xy} can be found at any point within the element merely by specifying an (x, y) in local coordinates and evaluating $[\mathbf{B}]$ and then multiplying through in (5.11). Four locations could be selected making $[\mathbf{B}]$ a 12 × 12 and (5.11) would give M_x, M_y and M_{xy} at these four locations. In the program the four quarter points are used to output moment values (Fig. 5.3.)

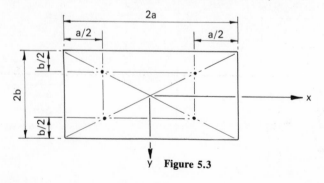

Figure 5.3

Table 5.1

$$[B]^T[D][B] = \frac{Et^3}{12(1-\nu^2)}\times$$

0	0	0	0	0	0	0	0	0	0	0	0
	0	0	0	0	0	0	0	0	0	0	0
		0	0	0	0	0	0	0	0	0	0
			$+4$	0	$12x$	$4v$	$4y$	$4xv$	$12yv$	$12vxy$	$12xyv$
				$2(1-v)$	0	0	$4x(1-v)$	$4y(1-v)$	0	$6x^2(1-v)$	$6y^2(1-v)$
					4	$12xv$	$4yv$	$4x$	$12y$	$12xyv$	$12xy$
						$36x^2$	$12xy$	$12x^2v$	$36xyv$	$36x^2y$	$36x^2yv$
							$4y^2+8x^2(1-v)$	$4xyv+8xy(1-v)$	$12y^2v$	$12xy^2+12x^3(1-v)$	$12xy^2-12xy^2(1-v)$
								$4x^2+8y^2(1-v)$	$12xy$	$12x^2yv+12x^2y(1-v)$	$12x^2y+12y^3(1-v)$
									$36y^2$	$36xy^2v$	$36xy^2$
										$36x^2y^2+18x^4(1-v)$	$36x^2y^2v+18x^2y^2(1-v)$
											$36x^2y^2+18y^4(1-v)$

SYMMETRICAL

(5.14)

Table 5.2

$$\frac{Et^3}{12(1-v^2)}\,.\,4ab.\int_{-a}^{a}\int_{-b}^{b}[\mathbf{B}]^{\mathrm{T}}[\mathbf{D}][\mathbf{B}]\,dxdy =$$

$$
\begin{bmatrix}
0 & 0 & 0 & 0 & 0 & 0 & 0 & 0 & 0 & 0 & 0 \\
  & 0 & 0 & 0 & 0 & 0 & 0 & 0 & 0 & 0 & 0 \\
  &   & 4 & 0 & 4v & 0 & 0 & 0 & 0 & 0 & 0 \\
  &   &   & 2(1-v) & 0 & 0 & 0 & 0 & 0 & 2(1-v)a^2 & 0 \\
  &   &   &   & 4 & 0 & 0 & 4va^2 & 0 & 0 & 2(1-v)b^2 \\
  &   &   &   &   & 12a^2 & 0 & 0 & 0 & 0 & 0 \\
  &   &   &   &   &   & \tfrac{4}{3}.b^2+\tfrac{8}{3}.ba^2(1-v) & 0 & 4b^2v & 0 & 0 \\
  &   &   &   &   &   &   & \tfrac{4}{3}a^2+\tfrac{8}{3}b^2(1-v) & 0 & 0 & 0 \\
  &   &   &   &   &   &   &   & 12b^2 & 0 & 0 \\
  &   &   &   &   &   &   &   &   & 4a^2b^2+\tfrac{18}{5}a^4(1-v) & 2(1+v)a^2b^2 \\
  &   &   &   &   &   &   &   &   &   & 4a^2b^2+\tfrac{18}{5}.(1-v)b^4
\end{bmatrix}
$$

SYMMETRICAL

(5.15)

Nodal loads. Unlike the previous programs the nodal load vector is progressively built up element by element, by multiplying the element stiffness matrix by the element displacement vector. This avoids having to assemble the uncondensed structure stiffness matrix.

Thus a complete set of nodal forces and moments ps, T_x's and T_y's is found which can be checked against the applied loads and moments which give the reaction values at boundary nodes.

5.2 Implementation in program

5.2.1 Description of program

The program **FEPB** is given in full in section 5.2.3 below. The data must be inserted into the program in DATA statements between lines 5000 and 10000. This form of data input means that the data is held within the program and can be stored on disc or tape with the program for future running and correction if required.

Stiffness matrices. Each element is dealt with in turn. Should the type be identical to a previous element, the stiffness matrix is as found for this previous element and the program skips to the point where the element stiffness coefficients are lodged into the structure condensed stiffness matrix.

Should the element be the first of a type then the following procedure is initiated. The matrix $\int_{-a}^{a} \int_{-b}^{b} [\mathbf{B}]^{-T} [\mathbf{D}] [\mathbf{B}]$ dy dx is established and stored as $[\mathbf{BDB}]$. Matrix $[\mathbf{A}]$ is then built up and its inverse $[\mathbf{A}]^{-1}$ found using a standard procedure. Finding the inverse involves much numerical manipulation and is thus very time-consuming. The inverse is stored in $[\mathbf{AH}]$ for use at the end of the program to find the element moments. The transpose of the inverse $[\mathbf{AT}]$ is multiplied into $[\mathbf{BDB}]$ and the result multiplied into the inverse is $[\mathbf{A}]$. This triple product is in fact the element stiffness matrix which is held as $[\mathbf{KL}]$ and stored for future use as $[\mathbf{KH}]$. As with the other programs $[\mathbf{KL}]$ must be incorporated into the main structure stiffness matrix $[\mathbf{KS}]$ which is similarly condensed. The restraints RE(I, 1), (I,2) and (I,3) are used as follows:

$$RE\,(I, 1) = 1 \text{ indicates w at node I is zero}$$
$$-\text{else} = 0$$

$$RE(I, 2) = 1 \text{ indicates } \theta_x \text{ at I is zero}$$
$$-\text{else} = 0$$

$$RE(I, 3) = 1 \text{ indicates } \theta_y \text{ at I is zero}$$
$$-\text{else} = 0$$

The 12 values of RE for each element indicate which rows and columns have to be transferred into $[KS]$, their locations being found as before from array MN which is evaluated at the beginning of the program. The final size of $[KS]$ is IN × IN.

Load vector. From the nodal force and moment data at each node (P, T_x, T_y), and uncondensed load vector is built up in $\{PH\}$, which is finally of size IN ×1.

Solution procedure. As with the previous programs, the same solution procedure is used to solve the equations

$$[KS]\{U\} = \{PH\}$$

the solution of unknown displacements and rotations being held in $\{PH\}$ before being expanded into $\{P\}$ by including all the boundary conditions of zero displacement and slope.

Element moments. Each element is dealt with in turn. From $\{P\}$, the 12 × 1 displacement vector for the element is extracted and held in $\{P1\}$. In order to find the element moments M_x, M_y and M_{xy}, the triple product $[D]$ $[B]$ $[AH]$ must be calculated and multiplied into $\{P1\}$. $[AH]$, the inverse of $[A]$, is retrieved from the store. However before these multiplications take place, the program calculates the element moments at the four quarter points in the element, the local coordinates of which are calculated in XL and YL. Using these, the first three rows of $[B]$, a 12 × 12 matrix, are calculated with XL and YL at the first quarter point, the next three rows (four to six) at the second point and so on. In the program the order of multiplication is $[AH]$ $\{P1\}$, which is then multiplied by $[B]$, before all being multiplied by $[D]$. The result of all the multiplication lands up again in a 12 × 1 vector $\{P\}$ which then holds M_x, M_y and M_{xy} at the four quarter points.

Note: the constant switching between $\{P\}$ and $\{P1\}$ during these multiplications is to save array storage.

Nodal forces. In order to find the nodal forces to check against applied loads and to find reaction forces and moments, a different approach is adopted here from the other three programs.

As each element is processed to find the moments, the element displacement vector $\{P1\}$ is multiplied by the element stiffness matrix $[KH]$ which is retrieved from store. This product results in the nodal loads P, T_x T_y at each of the elements 4 nodes being found – but these values are from the element alone and should be algebraically added to the values of P, T_x, T_y at the same node from elements which share the same node. Thus the values of P, T_x, T_y at each node are progressively summed as each element is processed, the results being accumulated in the vector $\{R\}$, the length of which is 3 times the number of nodes in the grid. At the end of the program the values in $\{R\}$ are printed out for each node.

At nodes of zero displacement w,(RE(I, 1) = 1) and/or zero rotation or slope θ_x (RE(I, 2) = 1) and/or zero rotation or slope θ_y (RE(I, 3) = 1), the corresponding values in $\{R\}$ will give the reaction P and/or T_x and/or T_y respectively. At nodes where there are applied loads and moments, the values in $\{R\}$ should equal these. Finally, of course, at 'free' nodes, the values in $\{R\}$ should be zero, subject to round-off errors etc.

5.2.2 Data preparation

A data preparation sheet is found in Appendix 5.1. The data is required in the following order, and where more than one item of data is required for entry on any line, a comma is used as a separator.

Structure data.

(i) Number of nodes in the grid, NN. This includes all boundary nodes.

(ii) Number of elements in the grid, NE.

Note that as written the program will solve for grids of up to 10 elements of 3 different types with a total of 20 unknown displacements and rotations (degrees of freedom) and 15 nodes. To change these limits EX, NX, TX and MX should be changed.

(iii) Plate properties: T – thickness
E – elastic modulus
NU – Poisson's ratio

Note: these refer to the whole plate, i.e. all the elements.

Element data. For each element starting with element ☐1️⃣,

(iv) Element node numbers – the four nodes defining the element should be given anticlockwise and the first node should be the one with the most positive x- and y- coordinates. Remember, as formulated, the sides of the element should be parallel to the structure or plate axes.

(v) Element type (TY): The very first element is type 1 and any subsequent elements of the same size should also be numbered as 1. Should the second element be different, it should be typed 2 and so on. Remember the program will only deal with 3 element types and the more elements of the same type, the quicker the program will run!

Nodal data. For each node, starting with node ①,

(vi) Coordinates x, y of the node, in *global or structural coordinates* – the program will calculate the local coordinates.

(vii)   Restraints, $RE(I, 1) = 1$ for $w_I = 0$, else $= 0$
$RE(I, 2) = 1$ for $\theta_{xI} = 0$, else $= 0$
$RE(I, 3) = 1$ for $\theta_{yI} = 0$, else $= 0$

(VIII)   Applied loads: P, z-direction load, transverse to plate surface (xy plane)
T_{ix} – torque about axis
T_y – torque about y-axis

Notes (a)   The element numbers and node numbers should start off with one and end with NE and NN respectively. This is implicitly assumed in the program. Remember the convention in element node numbering.

(b)   Units are as decided by the user, but must be consistent. For example if E is in units of N/mm^2, then the coordinates x, y and the thickness must be in mm and the applied forces must be in N and torques in Nmm. The output will have displacements in mm, slopes or rotations (θ) in radians, and moments and forces in Nmm and N respectively.

(c)   It is recommended that element and nodal data be carefully checked when printed out at the beginning of the output before allowing the program to proceed.

(d)   Elements should have their aspect ratios (length/breadth) as near unity as possible.

5.2.3   Program *FEPB*

Flow chart for **FEPB**

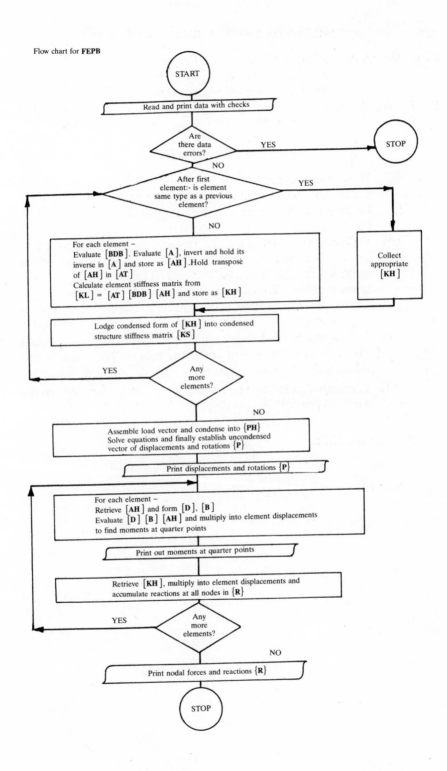

START

Read and print data with checks

Are there data errors? — YES → STOP

NO

After first element:- is element same type as a previous element? — YES

NO

For each element –
Evaluate $[\mathbf{BDB}]$. Evaluate $[\mathbf{A}]$, invert and hold its inverse in $[\mathbf{A}]$ .Hold transpose of $[\mathbf{AH}]$ in $[\mathbf{AT}]$
Calculate element stiffness matrix from
$[\mathbf{KL}] = [\mathbf{AT}] \, [\mathbf{BDB}] \, [\mathbf{AH}]$ and store as $[\mathbf{KH}]$

Collect appropriate $[\mathbf{KH}]$

Lodge condensed form of $[\mathbf{KH}]$ into condensed structure stiffness matrix $[\mathbf{KS}]$

YES — Any more elements?

NO

Assemble load vector and condense into $\{\mathbf{PH}\}$
Solve equations and finally establish uncondensed vector of displacements and rotations $\{\mathbf{P}\}$

Print displacements and rotations $\{\mathbf{P}\}$

For each element –
Retrieve $[\mathbf{AH}]$ and form $[\mathbf{D}]$, $[\mathbf{B}]$
Evaluate $[\mathbf{D}] \, [\mathbf{B}] \, [\mathbf{AH}]$ and multiply into element displacements to find moments at quarter points

Print out moments at quarter points

Retrieve $[\mathbf{KH}]$, multiply into element displacements and accumulate reactions at all nodes in $\{\mathbf{R}\}$

YES — Any more elements?

NO

Print nodal forces and reactions $\{\mathbf{R}\}$

STOP

```
100 REM              *** PROGRAM FEPB ***
105 REM            AUTHOR:
110 REM                    DAVID K. BROWN
112 REM         DEPARTMENT OF MECHANICAL ENGINEERING
115 REM                 UNIVERSITY OF GLASGOW
120 REM                      SCOTLAND
125 REM
126 REM                   AUGUST  1983
127 REM
130 REM    **   THE PROGRAM ALLOWS UP TO 10 ELEMENTS
131 REM         (OF 3 DIFFERENT TYPES)
132 REM         AND 15 NODES WITH A TOTAL OF
133 REM         20 DEGREES OF FREEDOM.               **
134 REM
135 REM    **   THE ELEMENTS ARE RECTANGULAR        **
136 REM
137 REM    **   FORCE AND MOMENT LOADS ARE APPLIED
138 REM         AT THE NODES AND THE BOUNDARIES ARE
139 REM         EITHER 'FIXED' OR 'FREE' (TO
140 REM         DISPLACE AND ROTATE).               **
145 REM
150 REM    **   ALL ELEMENTS ARE ASSUMED TO HAVE
155 REM         THE SAME THICKNESS, ELASTIC MODULUS
160 REM         AND POISSON'S RATIO.                **
165 REM
170 REM    **   INPUT TO THE PROGRAM IS THROUGH
171 REM         DATA STATEMENTS BETWEEN LINES
172 REM         5000   AND   10000.                 **
175 REM
180 REM    **   OUTPUT CONSISTS OF:
181 REM            NODAL DISPLACEMENTS AND ROTATIONS,
182 REM            ELEMENT MOMENTS MX, MY, MXY
183 REM            AT THE QUARTER POINTS,
184 REM            NODAL LOADS (WHICH INCLUDE
185 REM            REACTION FORCES AND TORQUES).     **
190 REM
200 REM    **   INITIALIZE PARAMETERS               **
210 REM
220 EX=10:NX=15:REM MAX # OF ELEMENTS AND MAX # OF NODES
230 TX=3:MX=20:REM MAX # OF TYPES AND MAX # OF DEGREES OF FREEDOM
240 PX=3*NX
250 DIM A(12,12),B(12,12),AT(12,12),BDB(12,12)
260 DIM KL(12,12),P(PX),XL(4),YL(4)
270 DIM N(EX,4),RE(NX,3),PH(PX),MN(NX,3)
280 DIM X(NX),Y(NX),KS(MX,MX),TY(EX),R(PX)
290 DIM AH(TX,12,12),D(3,3),P1(12),KH(TX,12,12)
310 FOR I= 1 TO 12
320 P1(I)=0
330 NEXT I
340 FOR I = 1 TO PX
350 PH(I) = 0:P(I)=0.:R(I)=0.
360 NEXTI
370 FOR I = 1 TO MX
380 FOR J = 1 TO MX
390 KS(I,J) = 0
400 NEXT J
410 NEXT I
420 REM    **   READ IN DATA FROM DATA STATEMENTS   **
```

```
430 READ NN,NE
440 IFNE>EX THEN PRINT "TOO MANY ELEMENTS...>",EX
450 IF NE>EX THEN 4310
460 IFNN>NX THEN PRINT "TOO MANY NODES...>",NX
470 IFNN>NX THEN 4310
480 GOSUB 51000 :REM OPEN PRINTER CHANNEL
481 GOSUB 53000 :REM MACHINE SPECIFIC STRINGS
482 READ T,E,NU
485 P$="RUN OF PROGRAM   F E P B"
486 GOSUB50000:IP=3:GOSUB50020
490 P$="          *** DATA INPUT ***"
491 GOSUB 50000:IP=2:GOSUB 50020
495 P$="NUMBER OF NODES     = "+STR$(NN):GOSUB50000
500 P$="NUMBER OF ELEMENTS  = "+STR$(NE):GOSUB50000
505 P$="ALL ELEMENTS HAVE THE SAME:-"
506 GOSUB50000:GOSUB50010
507 PB$="                              "
508 XS=T:GOSUB20000:P$=XS$
510 P$=PB$+"       THICKNESS      = "+P$
511 GOSUB 50000
514 XS=E:GOSUB10000:P$=XS$
515 P$=PB$+"       ELASTIC MODULUS = "+P$
516 GOSUB 50000
519 XS=NU:GOSUB20000:P$=XS$
520 P$=PB$+"   AND POISSON'S RATIO = "+P$
521 GOSUB 50000:IP=2:GOSUB 50020
530 FOR I = 1 TO NE
540 FOR J = 1 TO 4
550 READ N(I,J)
560 NEXT J
570 READ TY(I)
575 IFTY(I)>>TXTHENGOTO585
580 NEXT I
582 GOTO590
585 GOSUB 52000 :REM CLOSE PRINTER  CHANNEL
586 PRINT"ERROR - TYPE >";TX:GOTO4310
590 P$="ELEMENT    N O D E S        TYPE":GOSUB50000
600 P$="  NO     1  2  3  4":GOSUB50000:GOSUB50010
610 FORI=1TONE
620 P$="   "+STR$(I)+"       "+STR$(N(I,1))+" "+STR$(N(I,2))+" "
621 P$=P$+STR$(N(I,3))+" "+STR$(N(I,4))+"        "+STR$(TY(I))
622 GOSUB 50000
630 NEXT I:IP=2:GOSUB50020
640 FOR I = 1 TO NN
650 READ X(I),Y(I)
660 NEXT I
670 FOR I = 1 TO NN
680 READ RE(I,1),RE(I,2),RE(I,3)
690 READ PH(3*I-2),PH(3*I-1),PH(3*I)
700 NEXT I
710 P$="NODE       C O O R D I N A T E S        "
711 P$=P$+"R  E   S   T   R   A   I   N   T   S"
712 GOSUB 50000
720 P$="  NO        X           Y             "
721 P$=P$+"W       ANGLE X      ANGLE Y"
722 GOSUB 50000
725 GOSUB50010
730 FORI=1TONN:P$=STR$(I)+"       "
740 XS=X(I):GOSUB20000:P$=P$+XS$
750 XS=Y(I):GOSUB20000:P$=P$+XS$
```

```
760 P$=P$+"         "
765 FOR J=1 TO 3:P$=P$+STR$(RE(I,J))+"              ":NEXT J
770 GOSUB 50000:NEXT I
780 IP=2:GOSUB 50020
790 P$= "NODE        A P P L I E D    L O A D S"
795 GOSUB50000
800 P$=" NO           PZ          TX          TY"
801 GOSUB 50000:GOSUB 50010
805 FORI=1TONN:P$=""
810 FOR J = -2 TO 0:XS=PH(3*I+J):GOSUB 20000:P$=P$+XS$:NEXT J
820 P$=STR$(I)+"        "+P$:GOSUB50000
830 NEXTI:IP=2:GOSUB50020
835 GOSUB 52000 :REM CLOSE PRINTER  CHANNEL
840 PRINT"DO YOU WISH TO CORRECT THE DATA [Y/N]   ";N$;L4$;
845 INPUT AN$
850 IF LEFT$(AN$,1)="Y"THEN GOSUB 52000:GOTO 4310
860 IF LEFT$(AN$,1)<>"N" THENS40
870 PRINT "THE PROGRAM IS NOW RUNNING"
875 PRINT "(AND MAY TAKE SEVERAL MINUTES!!)
880 GOSUB 51000 :REM OPEN PRINTER CHANNEL
885 IN = 0
890 FOR I = 1TO NN
900 FOR J = 1 TO 3
910 IF RE(I,J) = 1 THEN 940
920 IN = IN+1
930 MN(I,J) = IN
940 NEXT J
950 NEXT I
960 MK = IN
970 P$="TOO MANY DEGREES OF FREEDOM"
975 IF MK>MX THEN GOSUB50000:GOTO 4310
980 REM  **  SCAN THROUGH EACH ELEMENT, FIND ITS
985 REM       STIFFNESS MATRIX AND LODGE IT INTO
990 REM       THE STRUCTURE STIFFNESS MATRIX      **
1000 FOR NI = 1 TO NE
1002 T=TY(NI)
1003 IF NI = 1 THEN GOTO 1070
1004 FORI=1TONI-1:IFT=TY(I)THENGOTO1040
1005 NEXTI
1006 GOTO1070
1040 P$="ELEMENT"+STR$(NI)+" IS TYPE"+STR$(T)
1041 P$=P$+" - STIFFNESS MATRIX ALREADY FOUND"
1042 GOSUB50000
1050 GOTO 2620
1060 REM  **  INITIALIZE ARRAYS                   **
1070 FOR I = 1 TO12
1080 FOR J = 1 TO 12
1090 A(I,J) = 0
1100 B(I,J) = 0
1110 BDB(I,J) = 0
1120 AT(I,J) = 0
1130 KL(I,J) = 0
1140 NEXT J
1150 NEXT I
1160 REM  **  EVALUATE MATRIX [BDB],
1170 REM  **  FIND LOCAL COORDS OF ELEMENT NODES  **
1180 N1 = N(NI,1)
1190 N2 = N(NI,2)
1200 N3 = N(NI,3)
1210 N4 = N(NI,4)
```

```
1220 XM=(X(N1)+X(N2)+X(N3)+X(N4))/4
1230 YM=(Y(N1)+Y(N2)+Y(N3)+Y(N4))/4
1240 FOR I = 1 TO 4
1250 NL = N(NI,I)
1260 XL(I) = X(NL)-XM
1270 YL(I) = Y(NL)-YM
1280 NEXT I
1290 X1 = XL(1)
1300 X2 = XL(2)
1310 Y1 = YL(1)
1320 Y4 = YL(4)
1330 AL = ABS((X1-X2)/2)
1340 BL = ABS((Y1-Y4)/2)
1345 REM  **  ONLY 'BD' OF THE VARIABLE NAME 'BDB'
1346 REM       IS SIGNIFICANT IN 'PET' BASIC.      **
1350 BDB(4,4) = 4
1360 BDB(4,6) = 4*NU
1370 BDB(5,5) = 2*(1-NU)
1380 BDB(5,11) = 2*(1-NU)*AL↑2
1390 BDB(5,12) = 2*(1-NU)*BL↑2
1400 BDB(6,6)= 4
1410 BDB(7,7) = 12*AL↑2
1420 BDB(7,9) = 4*NU*AL↑2
1430 BDB(8,8) = 4*BL↑2/3+8*(1-NU)*AL↑2/3
1440 BDB(8,10) = 4*NU*BL↑2
1450 BDB(9,9) = 4*AL↑2/3+8*(1-NU)*BL↑2/3
1460 BDB(10,10) = 12*BL↑2
1470 BDB(11,11) = 4*AL↑2*BL↑2+18*(1-NU)*AL↑4/5
1480 BDB(11,12) = 2*(1+NU)*AL↑2*BL↑2
1490 BDB(12,12) = 4*AL↑2*BL↑2+18*(1-NU)*BL↑4/5
1500 BDB(6,4) = BDB(4,6)
1510 BDB(11,5) = BDB(5,11)
1520 BDB(12,5) = BDB(5,12)
1530 BDB(9,7) = BDB(7,9)
1540 BDB(10,8) = BDB(8,10)
1550 BDB(12,11) = BDB(11,12)
1560 MLT = E*T↑3*4*AL*BL/12/(1-NU↑2)
1570 FOR I = 1 TO 12
1580 FOR J = 1 TO 12
1590 BDB(I,J) = MLT*BDB(I,J)
1600 NEXT J
1610 NEXT I
1620 REM  **  EVALUATE MATRIX [A]                **
1630 FOR II = 1 TO 4
1640 I = 3*II-2
1650 XI = XL(II)
1660 YI = YL(II)
1670 A(I,1) = 1
1680 A(I,2) = XI
1690 A(I,3) = YI
1700 A(I,4) = XI↑2
1710 A(I,5) = XI*YI
1720 A(I,6) = YI↑2
1730 A(I,7) = XI↑3
1740 A(I,8) = XI↑2*YI
1750 A(I,9) = XI*YI↑2
1760 A(I,10) = YI↑3
1770 A(I,11) = XI↑3*YI
1780 A(I,12) = XI*YI↑3
1790 I = I+1
```

```
1800 A(I,2) = -1
1810 A(I,4) = -2*XI
1820 A(I,5) = -YI
1830 A(I,7) = -3*XI↑2
1840 A(I,8) = -2*XI*YI
1850 A(I,9) = -YI↑2
1860 A(I,11) = -3*XI↑2*YI
1870 A(I,12) = -YI↑3
1880 I = I+1
1890 A(I,3) = 1
1900 A(I,5) = XI
1910 A(I,6) = 2*YI
1920 A(I,8) = XI↑2
1930 A(I,9) = 2*XI*YI
1940 A(I,10) = 3*YI↑2
1950 A(I,11) = XI↑3
1960 A(I,12) = 3*XI*YI↑2
1970 NEXT  II
1990 FOR I = 1 TO 12
2000 B(I,I)=1
2010 NEXT I
2020 REM  **  FIND THE INVERSE OF [A] AND LOCATE
2025 REM      IT BACK IN [A]. STORE IT IN [AH].    **
2030 P$="MATRIX INVERSION"
2035 GOSUB50000:IP=2:GOSUB50020
2040 FOR J = 1 TO 12
2050 FORI=JTO12
2060 IF A(I,J) <> 0 THEN 2100
2070 NEXTI
2080 GOSUB 52000 :REM CLOSE PRINTER  CHANNEL
2085 PRINT"A IS A SINGULAR MARTIX"
2090 GOTO 4310
2100 FOR K = 1 TO 12
2110 S = A(J,K)
2120 A(J,K) = A(I,K)
2130 A(I,K) = S
2140 S = B(J,K)
2150 B(J,K) = B(I,K)
2160 B(I,K) = S
2170 NEXT K
2180 TT = 1/A(J,J)
2190 FOR K = 1 TO 12
2200 A(J,K) = TT*A(J,K)
2210 B(J,K) = TT*B(J,K)
2220 NEXT K
2230 FOR L = 1 TO 12
2240 IF L=J THEN 2300
2250 TT= -A(L,J)
2260 FOR K = 1 TO 12
2270 A(L,K) = A(L,K)+TT*A(J,K)
2280 B(L,K) = B(L,K)+TT*B(J,K)
2290 NEXT K
2300 NEXT L
2310 NEXT J
2320 REM  **  TRANSPOSE THE INVERSE OF [A]
2325 REM      AND STORE AS [AT].              **
2330 FOR I = 1 TO 12
2340 FOR J = 1 TO 12
2350 AT(J,I) = B(I,J)
2360 A(I,J) = B(I,J)
```

```
2370 AH(T,I,J)=A(I,J)
2380 NEXT J
2390 NEXT I
2400 REM   **  EVALUATE TRIPLE PRODUCT [AT][BDB][A]
2405 REM       AND THUS FIND ELEMENT STIFFNESS
2410 REM       MATRIX [KL].  STORE AS [KH]          **
2420 FOR I = 1 TO12
2430 FOR J = 1 TO 12
2440 S = 0
2450 FOR K = 1 TO 12
2460 S = S+AT(I,K)*BDB(K,J)
2470 NEXT K
2480 B(I,J) = S
2490 NEXT J
2500 NEXT I
2510 FOR I = 1 TO12
2520 FOR J = 1 TO 12
2530 S = 0
2540 FOR K = 1 TO 12
2550 S = S+B(I,K)*A(K,J)
2560 NEXTK
2570 KH(T,I,J)= S
2580 KL(I,J) = S
2590 NEXT J
2600 NEXTI
2610 REM   **  LODGE [KL] INTO THE STRUCTURAL
2615 REM       STIFFNESS MATRIX [KS].              **
2620 FOR I = 1 TO 4
2630 ND = N(NI,I)
2640 IS = 3*I-2
2650 FOR L = 1 TO 3
2660 IF RE(ND,L)=1 THEN GOTO 2770
2670 PK = MN(ND,L)
2680 FOR J = 1 TO 4
2690 JS = 3*J-2
2700 NC = N(NI,J)
2710 FOR M = 1 TO 3
2720 IF RE(NC,M) = 1 THEN GOTO 2750
2730 PL = MN(NC,M)
2740 KS(PK,PL) = KS(PK,PL)+KH(T,IS+L-1,JS+M-1)
2750 NEXT M
2760 NEXT J
2770 NEXT L
2780 NEXT I
2790 NEXT NI
2795 REM   **  EVALUATE LOAD VECTOR AND
2800 REM       CONDENSE INTO [PH].                 **
2810 OT = 0
2820 FOR I = 1 TO NN
2830 FOR J = 1TO 3
2840 IFRE(I,J)=0 THEN GOTO 2890
2850 FOR M = 3*I-(3-J)-OT TO 3*NN-OT-1
2860 PH(M) = PH(M+1)
2870 NEXT M
2880 OT = OT+1
2890 NEXT J
2900 NEXT I
2910 M = 3*NN-OT
2920 M1 = 3*NN-OT-1
2930 FOR I = 1 TO M1
```

```
2940 L=I+1
2950 FOR J=L TO M
2960 IF KS(J,I) = 0 THEN GOTO 3010
2970 FOR KK = L TO M
2980 KS(J,KK)=KS(J,KK)-KS(I,KK)*KS(J,I)/KS(I,I)
2990 NEXT KK
3000 PH(J) = PH(J)-PH(I)*KS(J,I)/KS(I,I)
3010 NEXT J
3020 NEXT I
3030 PH(M) = PH(M)/KS(M,M)
3040 FOR I = 1 TO M1
3050 KK = M-I
3060 L = KK+1
3070 FOR J = L TO M
3080 PH(KK) = PH(KK)-PH(J)*KS(KK,J)
3090 NEXT J
3100 PH(KK) = PH(KK)/KS(KK,KK)
3110 NEXT I
3120 IN = 0
3130 FOR I = 1 TO NN
3140 FOR J = 1 TO 3
3150 IF RE(I,J)=1 THEN GOTO 3180
3160 IN = IN+1
3170 P(3*I-3+J)= PH(IN)
3180 NEXT J
3190 NEXT I
3200 REM  **   PRINT OUT RESULTS                    **
3210 IP=2:GOSUB50020:
3211 P$="        *** OUTPUT RESULTS ***"
3212 GOSUB 50000:IP=2:GOSUB 50020
3220 P$="* NODAL DISPLACEMENTS AND ROTATIONS *"
3221 GOSUB 50000:GOSUB 50010
3230 P$="NODE DISPLACEMENT   R O T A T I O N S"
3231 GOSUB 50000
3240 P$=" NO          W          ANGLE X      ANGLE Y"
3241 GOSUB 50000:GOSUB 50010
3250 FOR I = 1 TO NN
3260 P$=""
3270 FOR J = 1 TO3
3280 XS=P(3*I-3+J):GOSUB20000:P$=P$+XS$
3290 NEXT J
3295 P$=STR$(I)+"   "+P$:GOSUB50000
3310 NEXT I
3320 REM  **   SET UP PRODUCT [D][B][INVERSE OF A]
3325 REM      TO SOLVE FOR THE ELEMENT MOMENTS
3330 REM      MX, MY, MXY AT THE QUARTER POINTS.  **
3340 FOR I = 1 TO 3*NN
3350 PH(I) = P(I)
3360 NEXT I
3370 MLT=E*T↑3/12/(1-NU↑2)
3380 D(1,1)=MLT
3390 D(1,2)=-NU*MLT
3400 D(1,3)=0
3410 D(2,1)=-NU*MLT
3420 D(2,2)=MLT
3430 D(2,3)=0
3440 D(3,1)=0
3450 D(3,2)=0
3460 D(3,3)=MLT*(1-NU)/2
3470 FORI=1TO45:R(I)=0:NEXTI
```

```
3475 IP=2:GOSUB50020
3476 P$="* MOMENTS AT QUARTER POINTS *"
3477 GOSUB 50000:GOSUB 50010
3480 P$="ELEMENT        LOCATION                    "
3481 P$=P$+"M  O  M  E  N  T  S"
3482 GOSUB50000
3490 P$="  NO        X              Y              "
3491 P$=P$+"MX        MY            MXY"
3492 GOSUB50000
3500 FOR NI= 1 TO NE
3510 FOR I=1 TO 12
3520 P1(I) = 0
3530 P(I) = 0
3540 FOR J=1 TO 12
3550 B(I,J) =0
3560 NEXT J
3570 NEXTI
3580 REM  **  FIND QUARTER POINT COORDINATES
3590 N1 = N(NI,1)
3600 N2 = N(NI,2)
3610 N3 = N(NI,3)
3620 N4 = N(NI,4)
3630 XM=(X(N1)+X(N2)+X(N3)+X(N4))/4
3640 YM=(Y(N1)+Y(N2)+Y(N3)+Y(N4))/4
3650 FOR I = 1 TO 4
3660 NL = N(NI,I)
3670 XL(I) =(X(NL)-XM)/2
3680 YL(I) =(Y(NL)-YM)/2
3690 NEXT I
3700 FOR I= 1 TO 4
3710 NB = N(NI,I)
3720 FOR J = 1 TO 3
3730 P1(3*I-3+J)=PH(3*NB-3+J)
3740 NEXT J
3750 NEXTI
3760 T=TY(NI)
3770 REM  **  PRODUCT [AH][DISPLACEMENTS] HELD IN [P] **
3780 FOR I=1 TO 12
3790 FOR J=1 TO 12
3800 P(I)=P(I)+AH(T,I,J)*P1(J)
3810 NEXT J:NEXTI
3820 FORI=1TO4:ND=N(NI,I):FORII=1TO3:FORJ=1TO12
3830 R(3*ND-3+II)=R(3*ND-3+II)+KH(T,3*I-3+II,J)*P1(J)
3840 NEXTJ:NEXTII:NEXTI
3850 REM  **  SET UP [B] AT QUARTER POINTS.      **
3860 FOR  K = 1 TO 12 STEP 3
3870 I = (K-1)/3+1
3880 XK =XL(I)
3890 YK=YL(I)
3900 B(K+1,6)=-2
3910 B(K+1,9)=-2*XK
3920 B(K+1,10)=-6*YK
3930 B(K+1,12)=-6*XK*YK
3940 B(K,4)=-2
3950 B(K,7)=-6*XK
3960 B(K,8)=-2*YK
3970 B(K,11)=-6*XK*YK
3980 B(K+2,5)=2
3990 B(K+2,8)=4*XK
4000 B(K+2,9)=4*YK
```

```
4010 B(K+2,11)=6*XK↑2
4020 B(K+2,12)=6*YK↑2
4030 NEXT K
4040 REM  **   PRODUCT [B][AH][DISP'TS] HELD IN [P1]   **
4050 FOR I= 1 TO 12
4060 FOR J = 1 TO 12
4070 P1(I)=P1(I)+B(I,J)*P(J)
4080 NEXT J
4090 NEXT I
4100 FOR I = 1 TO 12
4110 P(I)=0
4120 NEXT I
4130 REM  **   PRODUCT [D][B][AH][DISPLACEMENTS]
4135 REM          HELD IN [P] GIVES ELEMENT MOMENTS. **
4140 FOR I = 1 TO 4
4150 FOR J = 1 TO 3
4160 FOR K = 1 TO 3
4170 P(3*I-3+J)=P(3*I-3+J)+D(J,K)*P1(3*I-3+K)
4180 NEXT K
4190 NEXT J
4200 NEXT I
4210 FORI=1TO4:XL(I)=XL(I)+XM:YL(I)=YL(I)+YM:NEXTI
4220 FORI=1TO4:II=I*3-2
4225 XS=XL(I):GOSUB 20000:P$=XS$
4226 XS=YL(I):GOSUB 20000:P$=P$+XS$
4227 XS=P(II):GOSUB 20000:P$=P$+XS$
4228 XS=P(II+1):GOSUB 20000:P$=P$+XS$
4229 XS=P(II+3):GOSUB 20000:P$=P$+XS$
4230 P$=STR$(NI)+"     "+P$:GOSUB 50000
4240 NEXTI:GOSUB 50010
4250 NEXT NI
4260 IP=2:GOSUB 50020
4261 P$="* NODAL FORCES * "
4262 GOSUB 50000:GOSUB 50010
4270 P$="NODE        N O D A L    F O R C E S"
4271 GOSUB 50000
4280 P$=" NO           P          TX           TY"
4281 GOSUB 50000
4290 FW=15:NS=3:FORJ=1TONN:P$=""
4292 FORI=-2TO0
4293 XS=R(3*J+I):GOSUB 20040:P$=P$+XS$
4294 NEXT I
4300 P$=STR$(J)+"     "+P$:GOSUB 50000
4305 NEXT J:IP=3:GOSUB 50020
4310 P$="*****    END OF RUN - PROGRAM FEPB *****"
4311 GOSUB50000:IP=5:GOSUB50020
4315 END
5000 REM *******************************
5010 REM DATA STATEMENTS LOCATED BETWEEN
5020 REM LINES 5000 AND 10000
5030 REM *******************************
10000 REM             FORMATTING AND INPUT/OUTPUT
10001 REM                 SUBROUTINES BY
10002 REM                 DAVID A. PIRIE
10003 REM     DEPARTMENT OF AERONAUTICS & FLUID MECHANICS
10004 REM             UNIVERSITY OF GLASGOW
10005 REM                   SCOTLAND
10006 REM                 AUGUST  1983
10010 REM
10015 REM  **  FORMAT NUMERICAL OUTPUT IN
```

```
10020 REM         SCIENTIFIC NOTATION                    **
10035 FW=12:NS=4
10040 WE =1E-30
10045 KE=0:KE$="":BL$="              ":B0$="00000000"
10050 F5=FW-NS-5:N3=NS+3:Z$="0.":AX=ABS(XS)
10052 IF AX<WE THEN XS$=LEFT$(BL$,F5)+Z$+LEFT$(BL$,N3):GOTO10095
10055 IFABS(XS)<.01ORABS(XS)>=1E9THEN10080
10060 IFABS(XS)<10ORABS(XS)>=10THENGOSUB10175
10065 GOSUB10110
10070 GOTO10095
10080 XS$=STR$(XS):KE$=RIGHT$(XS$,3):KE=VAL(KE$)
10085 XS=VAL(LEFT$(XS$,LEN(XS$)-4))
10090 GOSUB10110
10095 RETURN
10110 REM FORM O/P$
10115 GOSUB10145
10120 IFABS(XS)>=10THENGOSUB10175
10125 GOSUB10200
10130 GOSUB10225
10135 RETURN
10145 REM ROUNDOFF MANTISSA
10155 XR=5:FORI5=1TONS:XR=XR/10:NEXTI5
10160 XS=XS+XR*SGN(XS)
10165 RETURN
10175 REM NORMALISE MANTISSA
10180 IF ABS(XS)<1THENXS=XS*10:KE=KE-1:GOSUB10180
10185 IF ABS(XS)>=10THENXS=XS/10:KE=KE+1:GOSUB10185
10190 RETURN
10200 REM FORM EXPONENT$
10205 S$="+":IFKE<0THENS$="-"
10210 KE$=S$+RIGHT$("0"+MID$(STR$(KE),2),2)
10215 RETURN
10225 REM FORM (MANT+EXP)$
10230 X1$=LEFT$(STR$(XS),NS+2)
10235 XS$=X1$+LEFT$(B0$,NS+2-LEN(X1$))
10240 IFXS=INT(XS)THEN XS$=X1$+"."+LEFT$(B0$,NS-1)
10245 XS$=LEFT$(BL$,FW-NS-6)+XS$+"E"+KE$
10250 RETURN
20000 REM   **  FORMAT NUMERICAL OUTPUT                  **
20020 FW=12:NS=3
20040 BL$="              "
20050 XS$=STR$(XS):XE$="    ":IFLEN(XS$)<4THENXS$=XS$+"    "
20060 IFABS(XS)>=10↑(8-NS)THENXX=XS:GOTO20080
20070 XX=XS+.5*SGN(XS)/10↑NS
20080 IFMID$(XS$,LEN(XS$)-3,1)="E"THENGOSUB20180
20090 XX$=STR$(XX)
20100 FORJ5=1TOLEN(XX$)
20110 IFMID$(XX$,J5,1)="."THENDP=J5:GOTO20130
20120 NEXTJ5:DP=LEN(XX$)+1:XX$=XX$+".0000000"
20130 XS$=LEFT$(XX$,DP+NS)+XE$
20140 LX=LEN(XS$):IFLX>FWTHENXS$=LEFT$(XS$,FW):GOTO20160
20150 XS$=LEFT$(BL$,FW-LX)+XS$
20160 RETURN
20180 XE$=RIGHT$(XS$,4):XR=VAL(RIGHT$(XS$,2))
20190 XX=VAL(LEFT$(XS$,LEN(XS$)-4))+.5*SGN(XS)/10↑NS
20200 RETURN
50000 REM ** THE FOLLOWING STATEMENTS
50001 REM     MUST BE TAILORED TO THE
50002 REM     PARTICULAR MACHINE IN USE              **
50005 PRINT#5,P$:RETURN :REM ** PRINTLINE ON PRINTER **
```

F

```
50010 PRINT#5:RETURN     :REM ** 1 LINEFEED ON PRINTER **
50015 REM
50020 FOR KP = 1 TO IP  :REM IP
50021 PRINT#5            :REM   LINEFEEDS
50022 NEXT KP            :REM    ON
50023 RETURN            :REM    PRINTER *
50025 REM
51000 OPEN 5,4:RETURN    :REM ** OPEN CHANNEL TO PRINTER  **
52000 CLOSE 5:RETURN     :REM ** CLOSE CHANNEL TO PRINTER **
53000 REM ** THE FOLLOWING Y$,N$,L4$,AK$
53001 REM      ARE USED WITH 'INPUT' STATEMENTS -
53002 REM      SET THEM ALL EQUAL TO "" IF
53003 REM      L4$ NOT POSSIBLE ON YOUR MACHINE      **
53010 Y$="Y ":N$="N ":AK$="* "
53020 L4$="▉▊▋▌":RETURN :REM ** L4$ = 4 'CURSOR-LEFTS'  **
READY.
```

5.2.4   *Example of solution to thin rectangular plate problem using **FEPB***

(a)   *The structure.* A square plate with all edges clamped and carrying a central load.

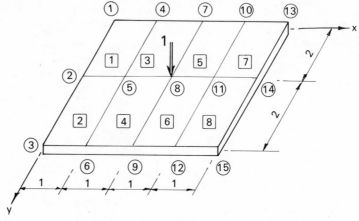

all edges clamped   t = 1   E = 1 $v = 0.3$

(b)   *The data.*

```
5000 DATA15.8                5160 DATA0,0,0,0,0,0
5010 DATA1,1,.3              5170 DATA1,1,1,0,0,0
5020 DATA5,2,1,4,1           5180 DATA1,1,1,0,0,0
5030 DATA6,3,2,5,1           5190 DATA0,0,0,1,0,0
5040 DATA8,5,4,7,1           5200 DATA1,1,1,0,0,0
5050 DATA9,6,5,8,1           5210 DATA1,1,1,0,0,0
5060 DATA11,8,7,10,1         5220 DATA0,0,0,0,0,0
5070 DATA12,9,8,11,1         5230 DATA1,1,1,0,0,0
5080 DATA14,11,10,13,1       5240 DATA1,1,1,0,0,0
5090 DATA15,12,11,14,1       5250 DATA1,1,1,0,0,0
5100 DATA0,0,0,2,0,4,1,0,1,2,1,4,2,0,2,2,2,4    5260 DATA1,1,1,0,0,0
5110 DATA3,0,3,2,3,4,4,0,4,2,4,4
5120 DATA1,1,1,0,0,0
5130 DATA1,1,1,0,0,0
5140 DATA1,1,1,0,0,0
5150 DATA1,1,1,0,0,0
```

(c)   *The solution.*

```
RUN OF PROGRAM   F E P B


              *** DATA INPUT ***


NUMBER OF NODES     =   15
NUMBER OF ELEMENTS  =   8
ALL ELEMENTS HAVE THE SAME:-

                                  THICKNESS         =    1.000
                                  ELASTIC MODULUS   =    1.000E+00
                            AND POISSON'S RATIO     =    .300


ELEMENT    N O D E S          TYPE
  NO       1  2  3  4

   1       5  2  1  4        1
   2       6  3  2  5        1
   3       8  5  4  7        1
   4       9  6  5  8        1
   5      11  8  7 10           1
   6      12  9  8 11           1
   7      14 11 10 13             1
   8      15 12 11 14             1


NODE     C O O R D I N A T E S      R   E   S   T   R   A   I   N   T  S
 NO         X            Y          W          ANGLE X        ANGLE Y

  1      0.000        0.000         1           1              1
  2      0.000        2.000         1           1              1
  3      0.000        4.000         1           1              1
  4      1.000        0.000         1           1              1
  5      1.000        2.000         0           0              0
  6      1.000        4.000         1           1              1
  7      2.000        0.000         1           1              1
  8      2.000        2.000         0           0              0
  9      2.000        4.000         1           1              1
 10      3.000        0.000         1           1              1
 11      3.000        2.000         0           0              0
 12      3.000        4.000         1           1              1
 13      4.000        0.000         1           1              1
 14      4.000        2.000         1           1              1
 15      4.000        4.000         1           1              1


NODE       A P P L I E D   L O A D S
 NO        PZ           TX           TY

  1       0.000        0.000        0.000
  2       0.000        0.000        0.000
  3       0.000        0.000        0.000
  4       0.000        0.000        0.000
  5       0.000        0.000        0.000
  6       0.000        0.000        0.000
  7       0.000        0.000        0.000
  8       1.000        0.000        0.000
```

```
9          0.000          0.000          0.000
10         0.000          0.000          0.000
11         0.000          0.000          0.000
12         0.000          0.000          0.000
13         0.000          0.000          0.000
14         0.000          0.000          0.000
15         0.000          0.000          0.000
```

```
MATRIX INVERSION

ELEMENT 2 IS TYPE 1 - STIFFNESS MATRIX ALREADY FOUND
ELEMENT 3 IS TYPE 1 - STIFFNESS MATRIX ALREADY FOUND
ELEMENT 4 IS TYPE 1 - STIFFNESS MATRIX ALREADY FOUND
ELEMENT 5 IS TYPE 1 - STIFFNESS MATRIX ALREADY FOUND
ELEMENT 6 IS TYPE 1 - STIFFNESS MATRIX ALREADY FOUND
ELEMENT 7 IS TYPE 1 - STIFFNESS MATRIX ALREADY FOUND
ELEMENT 8 IS TYPE 1 - STIFFNESS MATRIX ALREADY FOUND
```

```
        *** OUTPUT RESULTS ***


* NODAL DISPLACEMENTS AND ROTATIONS *

NODE DISPLACEMENT    R O T A T I O N S
NO      W          ANGLE X        ANGLE Y

1       0.000        0.000          0.000
2       0.000        0.000          0.000
3       0.000        0.000          0.000
4       0.000        0.000          0.000
5        .483        -.771          0.000
6       0.000        0.000          0.000
7       0.000        0.000          0.000
8       1.046       -1.252E-10      0.000
9       0.000        0.000          0.000
10      0.000        0.000          0.000
11       .483         .771          0.000
12      0.000        0.000          0.000
13      0.000        0.000          0.000
14      0.000        0.000          0.000
15      0.000        0.000          0.000
```

```
* MOMENTS AT QUARTER POINTS *

ELEMENT     LOCATION                M O M E N T S
  NO      X         Y          MX         MY          MXY
  1     .750      1.500       .025      -.049        -.075
  1     .250      1.500      -.075       .030        -.022
  1     .250       .500      -.022     -9.900E-04   -3.513E-03
  1     .750       .500    -3.513E-03   -.022         .483

  2     .750      3.500    -3.513E-03   -.022        -.022
  2     .250      3.500      -.022     -9.900E-04    -.075
  2     .250      2.500      -.075       .030         .025
  2     .750      2.500       .025      -.049         .483

  3    1.750      1.500       .167     6.573E-03      .069
  3    1.250      1.500       .069      -.046         .018
  3    1.250       .500       .018      -.044         .048
  3    1.750       .500       .048      -.071         .483

  4    1.750      3.500       .048      -.071         .018
  4    1.250      3.500       .018      -.044         .069
  4    1.250      2.500       .069      -.046         .167
  4    1.750      2.500       .167     6.573E-03      .483
```

```
5       2.750       1.500       .027        .095        .167
5       2.250       1.500       .167        6.573E-03   .048
5       2.250        .500       .048       -.071        .018
5       2.750        .500       .018       -.044        .483

6       2.750       3.500       .018       -.044        .048
6       2.250       3.500       .048       -.071        .167
6       2.250       2.500       .167        6.573E-03   .027
6       2.750       2.500       .027        .095        .483

7       3.750       1.500      -.075        .030       -.017
7       3.250       1.500      -.017        .092       -3.513E-03
7       3.250        .500      -3.513E-03  -.022       -.022
7       3.750        .500      -.022       -9.900E-04   .483

8       3.750       3.500      -.022       -9.900E-04  -3.513E-03 .
8       3.250       3.500      -3.513E-03  -.022       -.017
8       3.250       2.500      -.017        .092       -.075
8       3.750       2.500      -.075        .030        .483
```

```
* NODAL FORCES *

NODE            N  O  D  A  L    F  O  R  C  E  S
NO              P               TX              TY
1       4.971E-03        .037           -4.864E-03
2       -.224            .175            2.805E-12
3       4.971E-03        .037            4.864E-03
4       -.088            9.716E-03      -.067
5       -2.262E-10      -1.687E-10       8.356E-12
6       -.088            9.716E-03       .067
7       -.110           -1.421E-11      -.132
8       1.000           -4.809E-11      -6.983E-12
9       -.110            1.853E-11       .132
10      -.088           -9.716E-03      -.067
11      0.000            1.323E-10      -1.097E-11
12      -.088           -9.716E-03       .067
13      4.971E-03       -.037           -4.864E-03
14      -.224           -.175           -1.373E-12
15      4.971E-03       -.037            4.864E-03
```

```
*****    END OF RUN - PROGRAM FEPB *****
```

G

Appendix 5.1   FEPB (Computer program for thin flat plate analysis): program summary and data sheet

1. *Introduction*

Element and nodal data are input along with the plate properties – thickness elastic modulus and Poisson's ratio. For each element, the element type is checked against any previous element processed, and, if no element of this type has had its stiffness matrix calculated, the following process is followed. First the 12 × 12 matrix

$$\int_{-a}^{a} \int_{-b}^{b} [\mathbf{B}]^{\mathrm{T}} [\mathbf{D}] [\mathbf{B}] \, dy dx$$

is calculated and stored as $[\mathbf{BDB}]$ and then the matrix $[\mathbf{A}]$ is calculated. The matrix $[\mathbf{A}]$ inverted to $[\mathbf{A}]^{-1}$ which is held as $[\mathbf{AH}]$ before being further transposed to $[\mathbf{A}]^{-\mathrm{T}}$. The triple product

$$[\mathbf{A}]^{-\mathrm{T}} [\mathbf{BDB}] [\mathbf{A}]^{-1}$$

gives the element stiffness matrix, which is called $[\mathbf{KL}]$ in the program and is stored as $[\mathbf{KH}]$. The stiffness coefficients are then transposed into the structure condensed stiffness matrix $[\mathbf{KS}]$ depending on the zero displacement boundary conditions.

The load vector is uncondensed as read in, and must be condensed and stored as $\{\mathbf{PH}\}$.

The solution of the resulting equations gives all the unknown displacements.

For each element the product $[\mathbf{D}] [\mathbf{B}] [\mathbf{AH}]$ is multiplied into the element displacement vector to give the moments M_x, M_y and M_{xy} at the quarter points in the element. The same displacement vector is multiplied with the retrieved element stiffness matrix $[\mathbf{KH}]$ to give a set of element nodal forces which when algebraically summed with those from other elements will give a complete set of nodal forces $\{\mathbf{R}\}$. From $\{\mathbf{P}\}$ can be found the reaction forces and moments at the zero displacement and/or slope boundaries.

Preparation of input data for this program should be accomplished in the following sequence:
(1) Sketch the plate and discretize into a satisfactory number of rectangular elements, trying to keep as many as possible the same size and orientation.
(2) Number the nodes and elements.
(3) Establish a reference coordinate system and determine nodal coordinates. Remember that the element sides should be parallel to these axes.

(4) Remember the element nodes should be fed into the program in an anti-clockwise sense with the node with the most positive coordinates first, as shown.

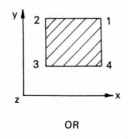

OR

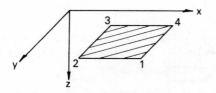

(5) Define the different load cases to be considered.
(6) Fill out the datasheet.

2. *Datasheet for FEPB*

(Note: units must be consistent.)

Structure data. Number of nodes (NN ⩽ 15), number of elements (NE ⩽10).

Plate properties:
thickness (T), elastic modulus (E), Poisson's ratio (NU).

	,	

	,	,	

Element data. For each element starting with ☐1 and ending with NE,

Node numbers[1]				Type[2]
,	,	,	,	,
,	,	,	,	,
,	,	,	,	,
,	,	,	,	,
,	,	,	,	,
,	,	,	,	,
,	,	,	,	,
,	,	,	,	,

Note: [1]Nodes numbered anticlockwise, node with most positive coordinates first.

[2]A maximum of three types is permitted.

Nodal data.
Coordinates for each node starting with ① and ending with NN.

X	Y
,	
,	
,	
,	
,	
,	
,	
,	
,	
,	

Restraints (=1 for zero displacement or slope else = 0) and nodal loads for each node starting with ① and ending with NN.

Restraints on				Applied nodal loads		
w	θ_x	θ_y		P	T_x	T_y
,	,	,		,	,	
,	,	,		,	,	
,	,	,		,	,	
,	,	,		,	,	
,	,	,		,	,	
,	,	,		,	,	
,	,	,		,	,	
,	,	,		,	,	
,	,	,		,	,	
,	,	,		,	,	

Notes: (a) This is merely a sample blank datasheet for up to 8 elements and 10 nodes.

(b) The data should be typed into the program in DATA statements between line numbers 5000 and 10000.

6 Further developments of programs

6.0 Introduction

As was stated in Chapter 1, the programs presented in Chapters 2 to 5 are written in an expanded form for the sake of clarity. No attempt was made to optimize either storage or speed of execution. Essentially the programs were written for teaching purposes and no thought was given to their possible commercial use.

However it is hoped that when experience with the technique and the programs has been gained, changes can be made to the source and thus the programs developed for individual requirements on specific microcomputer systems. In many ways the best way to familiarize oneself with the programs is to 'open up' the source and insert extra print statements!

To assist such developments two Appendices A and B provide a summary of matrix operations and BASIC statements. Many definitive texts exist on both these topics and further serious developments will require reference to these.

Below are given some possible developments which the reader may wish to implement. The section 6.1.7 on disc storage is very much system-dependent and will require close study of specific system manuals.

6.1 Possible developments

6.1.1 *Units*

As has been stated many times in the text, the units used are entirely the responsibility of the user. Consistent units for input quantities will dictate the units of the output. However, statements could be inserted into the program such that the user would be required to state the units being used – N and mm or KN and m, etc. Subsequent print-outs of data and results could have all quantities titled with the appropriate units selected from a range of print statements.

This development can easily be carried out and can act as a constant reminder to users of the programs to have consistent units and to check their input data thoroughly.

6.1.2 *Increasing number of elements and nodes*

The programs as written will be accommodated in a microcomputer with 32K core of storage. Many micros now have greater store or the potential to add on more store and so the maximum number of elements and nodes can be increased. This can be easily done by changing only a few constants in the programs. The appropriate arrays are dynamically dimensioned.

(a) **PJFRAME**

Line 230 – change MX for maximum number of members
Line 240 – change NX for maximum number of nodes

(b) **PLFRAME**

Line 240 – change NX for maximum number of nodes
Line 250 – change MX for maximum number of members

(c) **FEP**

Line 180 – change EX for maximum number of elements
and NX for maximum number of nodes

(d) **FEPB**

Line 220 – change EX for maximum number of elements
and NX for maximum number of nodes
Line 230 – change TX for maximum number of element types
and MX for maximum number of degrees of freedom

6.1.3 *Data through INPUT statements*

Apart from **PJFRAME**, all other programs have data input with READ and DATA statements. This latter approach allows the data to be stored with the program on disk or tape and thus to be amended and re-run several times as is often required for people using finite element programs for the first time. In fact, multiple sets of data can be stored in DATA statements at the end of a program and by appropriately changing a group of line numbers, the copy of a particular set of data can be 'promoted' to the set for first solution. An additional advantage of READ over INPUT is that interrogative input requires much more programming, as can be seen with **PJFRAME**.

However, all the READ statements in **PLFRAME, FEP** and **FEPB** can be changed to INPUT and additional prints to the screen be inserted to request data. This can be useful since comments and warnings can simultaneously be flashed to the screen. One other major advantage of INPUT over READ is that if a BASIC compiler is used, the compiled version of the programs will run faster and it would make no sense to compile the DATA statements with the program!

6.1.4 *Plotting of grids*

Many errors are made in data input to finite element programs. Whereas many checks can be put into programs, one of the quickest and surest tests which can be applied is a visual check of the grid. No attempt has been made to incorporate graphical output of the finite element grids from the four programs in this book. Such output can be channelled to the screen or to a digital plotter, but many micro systems have neither facility available. In addition, graphics statements, like disc statements, are very system-dependent.

However, where possible, it is strongly recommended that grid plotting be implemented. It can be very simply achieved. Once the grid details have been fed in, the maximum and minimum dimensions in the x- and y- directions should be found and thus the 'window' determined. This window should be scaled to the dimensions of the screen or the plotter. Thereafter the pen should be moved to the first node of the first element and the outline of the element drawn. Each element should then be drawn in turn. The drawing of the triangle and quadrilateral elements will in fact result in common edges of adjacent elements being drawn twice, but this is essential for the check. Element and node numbers may also be drawn on to the grid.

6.1.5   *Multiple load cases*

Two situations arise:

(a) In **PJFRAME** because the data is through INPUT statements and to avoid re-keying in all structure data, the user is offered a re-run at the completion of any solution. Should the user not discover an input data error until the solution is complete, the opportunity exists to correct the input data and re-run. This device is equally well suited for running the solution to the same structure but with a different loading.

However it should be clearly noted that in **PJFRAME,** the solution is *ab initio* and that the program re-establishes the stiffness matrix $[KH]$. In finite element programs, building the stiffness matrix is time-consuming and its repetition should be avoided. In **PJFRAME** the time loss is not significant but in **FEPB,** the time to form the stiffness matrix for even small problems is noticeable.

Nevertheless if INPUT statements are implemented in **PLFRAME, FEP** and **FEPB** then the above loop should also be included to eliminate the re-keying in of all data.

(b) From the comments made above about the forming of the stiffness matrix, an alternative approach to multiple load cases is preferred. The inputting of load data should be delayed until the stiffness matrix is established and condensed. The establishing and condensing of the load vector will then precede the solution of the equations to give the unknown displacements and rotations and the subsequent calculation of forces, moments, stresses etc. The offer to re-run with different loads is then made, and if accepted the program should be looped back to the line where the loads are input.

Care should be taken about the state of the arrays, some of which should be re-initialized.

6.1.6 Efficient storage

Reference should be made to the Appendix A and section A.2.1. The programming of this is substantial.

6.1.7 The use of disc storage

The use of a disc as a method of temporary storage has not been incorporated into the programs. Unfortunately the control statements for disc reading and writing are very machine- or system-dependent and much care has been taken to keep the programs as interchangeable as possible. However the use of temporary or scratch files on discs make a lot of sense from the viewpoint of reducing the requirement for permanent storage and thus increasing the possible size of problems to be solved. The necessity for disc storage does not exist in **PJFRAME** but its implementation in the other three programs would increase their potential.

(a) *Program **PLFRAME***. As each member is processed its 6×6 global stiffness matrix $[KG]$ is generated and then stored in $[KH]$ which is dimensioned as $(10,6,6)$ and which allows for up to only 10 such members. Later in the program each member's 6×6 matrix is retrieved in turn to be (i) multiplied by the member's own displacement vector to give the member end forces and moments and (ii) incorporated into $[KS]$, the uncondensed structure stiffness matrix.

Alternatively, $[KH]$, could be eliminated by writing each member's 6×6 matrix to disc sequentially. Once the last member has been processed the disc file could be (i) re-wound and the member matrices read down to calculate the end forces and (ii) re-wound again and the member matrices read down again to be incorporated into the matrix $[KS]$.

(b) *Program **FEP***. During the processing of each element, the 3×6 array from $\frac{1}{2}$ Area $[B]$ $[D]$ is stored in $[ST]$ which is dimensioned as $ST(10,3,6)$. Similarly the element's 6×6 stiffness matrix $[KL]$ is stored in $[KH]$ which is similarly dimensioned. Subsequently each element is processed and the product of $[ST]$ and its displacement vector $\{P\}$ yields finally a complete set of element stresses. At the end of the program, each elements stiffness matrix from $[KH]$ is incorporated into $[KS]$ the uncondensed structure stiffness matrix, which, when multiplied with the uncondensed displacement vector, produces the nodal forces and reactions.

Alternatively, $[KH]$ and $[ST]$ can be eliminated by writing sequentially the $[ST]$ matrices onto one disc and the $[KL]$ matrices onto another disc. The disc files would then be rewound and the matrices read off as required. If only one disc is available, $[\mathbf{ST}]$ and $[\mathbf{KL}]$ would be written alternately and sequentially for each element onto the same file and finally the file rewound. When the values of $[\mathbf{ST}]$ are required, a dummy read would be inserted to read off the values of $[\mathbf{KL}]$ to move the reading head to the start of the next element's $[\mathbf{ST}]$. After a second rewind the values of $[\mathbf{KL}]$ could similarly be read off with a dummy read of $[\mathbf{ST}]$ values.

(c)   *Program* **FEPB.** Like **FEP** above, the program stores two matrices for each element – the inverse of $[\mathbf{A}]$ in $[\mathbf{AH}]$ and the stiffness matrix $[\mathbf{KL}]$ in $[\mathbf{KH}]$. Both $[\mathbf{AH}]$ and $[\mathbf{KH}]$ are dimensioned as (3,12,12). The allocation of an element type TY means that this program is not restricted to only 3 elements – merely 3 types! $[\mathbf{AH}]$ is retrieved later in the program to determine moments at quarter points and $[\mathbf{KH}]$ to determine the elements contribution to the nodal forces and reactions.

Alternatively, $[\mathbf{AH}]$ and $[\mathbf{KH}]$ can be eliminated and a disc system used. Unlike FEP above, the alternate writing of $[\mathbf{AH}]$ and $[\mathbf{KH}]$ for a single element type disc file is not matched by the requirement to read off $[\mathbf{AH}]$ and $[\mathbf{KH}]$ in the same order. The two 12×12 matrices are referred to a type and not an element number. For example, if the element numbers 1,2,3,4 ... have the type numbers 1,3,2,3 ... then the disc file must be accessed to find the first pair of matrices then the third, the second, the third, etc. Two approaches are possible: either (i) dummy read statements could allow the pointer to move to the start of the correct matrix pair, which are then read in and the file again rewound; or (ii) each pair of matrices are read in and form a separate file. When required, the element type will move the pointer using random access to the correct file, which can then be read sequentially.

Both these approaches are very much machine-dependent.

Appendix A   Matrix algebra

This appendix is not intended as a primer for matrix algebra. In the first half the definitions are followed by some of the simple matrix manipulations. In the second section, the specific properties of structural stiffness matrices and their more efficient storage and subsequent solution procedures are briefly discussed.

A.1   Matrix properties and manipulations

A.1.1   *Definitions*

A matrix is an orderly array of numbers, constants or variables, or a mixture of all three. Consider the following matrix $[\mathbf{A}]$:

$$
[\mathbf{A}] =
\begin{bmatrix}
a_{11} & a_{12} & a_{13} \cdots\cdots\cdots a_{1n} \\
a_{21} & a_{22} & a_{23} \cdots\cdots\cdots a_{2n} \\
\vdots & \vdots & \vdots \\
a_{ml} & a_{m2} & a_{m3} \qquad\quad a_{mn}
\end{bmatrix}
$$

It is m rows and n columns of quantities 'a', the subscript denoting its row and column location within the array. $[\mathbf{A}]$ is thus defined as being of order m$\times$ n and any quantity a_{ij} would be located in row i at column j. If m = n, the matrix is called square, and if m $\neq$ n, then rectangular. For the case of m or n equal to 1 the resulting single-row 1 $\times$ n or m $\times$1 matrices are called row and column vectors respectively.

e.g. $\{\mathbf{B}\} = \begin{bmatrix} 1 \\ 2 \\ 7 \end{bmatrix}$ is a column vector

and $\{\mathbf{C}\} = [3\ 5\ 4]$ is a row vector

whereas $[\mathbf{D}] =$

$$\begin{bmatrix} 1 & x^2 & 2x & 2x+1 \\ x+4 & 5 & x^3 & x \\ 2 & 1+3x^2 & 7x & 1 \end{bmatrix}$$

is a 3 × 4 matrix,

and, for example, $d_{23} = x^3$.

A.1.2 *Transpose*

Consider the particular matrix $[\mathbf{A}]$

$$[\mathbf{A}] = \begin{bmatrix} 1 & 2 & 1 & 0 \\ 0 & 1 & 1 & 3 \\ 1 & 2 & 1 & 4 \end{bmatrix}$$

If the rows and columns are interchanged, the 3 × 4 matrix $[\mathbf{A}]$ is transposed into a 4 × 3 matrix $[\mathbf{A}]^{\mathrm{T}}$

$$\text{Thus } [\mathbf{A}]^{\mathrm{T}} = \begin{bmatrix} 1 & 0 & 1 \\ 2 & 1 & 2 \\ 1 & 1 & 1 \\ 0 & 3 & 4 \end{bmatrix}$$

In other words all quantities a_{ij} transpose to a_{ji}. Also, a row vector will become a column vector, and vice versa.

In most structural problems the stiffness matrices are square and symmetric such that $a_{ij} = a_{ji}$ and in such a case $[\mathbf{A}] = [\mathbf{A}]^{\mathrm{T}}$.

A.1.3 *Addition and subtraction*

Matrices can only be added or subtracted if they are of the same order, and this is achieved merely by adding or subtracting the corresponding terms.

Thus

$$[\mathbf{A}] + [\mathbf{B}] = \begin{bmatrix} a_{11} & a_{12} \\ a_{21} & a_{22} \end{bmatrix} + \begin{bmatrix} b_{11} & b_{12} \\ b_{21} & b_{22} \end{bmatrix}$$

$$= \begin{bmatrix} (a_{11} + b_{11}) & (a_{12} + b_{12}) \\ (a_{21} + b_{21}) & (a_{22} + b_{22}) \end{bmatrix} = [\mathbf{C}]$$

or for any quantity $c_{ij} = a_{ij} + b_{ij}$.

Similarly, $[A] - [B] = [C]$ would be found from $c_{ij} = a_{ij} - b_{ij}$
and $3[A] + 2[B] = [C]$ from $c_{ij} = 3a_{ij} + 2b_{ij}$.

A.1.4   Multiplication

Two matrices can be multiplied together only if the number of columns in the
first is equal to the number of rows in the second. Thus an m × n matrix can be
multiplied with a n × q matrix to give an m × q matrix,

e.g.
$$
\begin{bmatrix} 1 & 2 & 1 & 0 \\ 0 & 1 & 1 & 3 \\ 1 & 2 & 1 & 5 \end{bmatrix}
\begin{bmatrix} 2 & 1 \\ 1 & 2 \\ 0 & 2 \\ -1 & 1 \end{bmatrix}
=
\begin{bmatrix} 4 & 7 \\ -2 & 7 \\ -1 & 12 \end{bmatrix}
$$

$$\quad\quad 3 \times 4 \quad\quad\quad\quad 4 \times 2 \quad\quad\quad\quad\quad 3 \times 2$$

If a matrix $[A]$ is multiplied into a matrix $[B]$, the terms in the resulting
matrix $[C]$ are obtained by taking the scalar product of each row of $[A]$ with
each column of $[B]$ such that

$$c_{ij} = \sum_{k=1}^{n} a_{ik} \cdot b_{kj}$$

A.1.5   Transpose of a product

It can be shown that the transpose of the product of $[A]$ $[B]$ is equal to the
product $[B]^T [A]^T$.

$$\text{i.e. } ([A] [B])^T = [B]^T [A]^T$$

A.1.6   Determinant

If $[A]$ is a 3 × 3 square matrix

$$
\begin{bmatrix} a_{11} & a_{12} & a_{13} \\ a_{21} & a_{22} & a_{23} \\ a_{31} & a_{32} & a_{33} \end{bmatrix}
$$

the determinant $|A|$ associated with $[A]$ is defined to be the number

$$|A| = a_{11} \begin{vmatrix} a_{22} & a_{23} \\ a_{32} & a_{33} \end{vmatrix} - a_{12} \begin{vmatrix} a_{21} & a_{23} \\ a_{31} & a_{33} \end{vmatrix} + a_{13} \begin{vmatrix} a_{21} & a_{22} \\ a_{31} & a_{32} \end{vmatrix}$$

where for any numbers a, b, c and d

$$\begin{vmatrix} a & b \\ c & d \end{vmatrix} = ad - bc$$

e.g. If $[\mathbf{A}] = \begin{bmatrix} 2 & 2 & 3 \\ 2 & 1 & 1 \\ 4 & 1 & 2 \end{bmatrix}$ then $|\mathbf{A}| = -4$

A.1.7 *Inverse of a matrix*

The inverse of a square matrix, $[\mathbf{A}]$, is denoted by $[\mathbf{A}]^{-1}$ and defined by

$$[\mathbf{A}]^{-1}[\mathbf{A}] = [\mathbf{I}]$$

$[\mathbf{I}]$ is the identity or unit matrix, which has ones on the diagonal and zeros elsewhere such that $[\mathbf{A}][\mathbf{I}] = [\mathbf{A}]$. If $[\mathbf{A}]$ is a 3 x 3 matrix then

$$[\mathbf{I}] = \begin{bmatrix} 1 & 0 & 0 \\ 0 & 1 & 0 \\ 0 & 0 & 1 \end{bmatrix}, \text{ also } 3 \times 3$$

The evaluation of the inverse, $[\mathbf{A}]^{-1}$, is long and tedious but can be achieved from

$$[\mathbf{A}]^{-1} = \frac{\text{adj}[\mathbf{A}]}{|\mathbf{A}|}$$

where $|\mathbf{A}|$ is the determinant of $[\mathbf{A}]$ (and must obviously not be zero) and adj $[\mathbf{A}]$ is the adjoint to $[\mathbf{A}]$.

Example $[\mathbf{A}] = \begin{bmatrix} 1 & 2 & 3 \\ 4 & 5 & 6 \\ 7 & 8 & 10 \end{bmatrix}$

$$|\mathbf{A}| = 1 \begin{vmatrix} 5 & 6 \\ 8 & 10 \end{vmatrix} - 2 \begin{vmatrix} 4 & 6 \\ 7 & 10 \end{vmatrix} + 3 \begin{vmatrix} 4 & 5 \\ 7 & 8 \end{vmatrix}$$

$$= +2 + 4 - 9 = -3$$

$$\text{adj}\left[\mathbf{A}\right] = \begin{bmatrix} + \begin{vmatrix} 5 & 6 \\ 8 & 10 \end{vmatrix} & - \begin{vmatrix} 4 & 6 \\ 7 & 10 \end{vmatrix} & + \begin{vmatrix} 4 & 5 \\ 7 & 8 \end{vmatrix} \\[2mm] - \begin{vmatrix} 2 & 3 \\ 8 & 10 \end{vmatrix} & + \begin{vmatrix} 1 & 3 \\ 7 & 10 \end{vmatrix} & - \begin{vmatrix} 1 & 2 \\ 7 & 8 \end{vmatrix} \\[2mm] + \begin{vmatrix} 2 & 3 \\ 5 & 6 \end{vmatrix} & - \begin{vmatrix} 1 & 3 \\ 4 & 6 \end{vmatrix} & + \begin{vmatrix} 1 & 2 \\ 4 & 5 \end{vmatrix} \end{bmatrix}^T$$

$$= \begin{bmatrix} 2 & +2 & -3 \\ +4 & -11 & +6 \\ -3 & +6 & -3 \end{bmatrix} = \begin{bmatrix} 2 & +4 & -3 \\ +2 & -11 & +6 \\ -3 & +6 & -3 \end{bmatrix}$$

$$\left[\mathbf{A}\right]^{-1} = \frac{\text{adj}\left[\mathbf{A}\right]}{\left|\mathbf{A}\right|} = \begin{bmatrix} -2/3 & -4/3 & +3/3 \\ -2/3 & +11/3 & -6/3 \\ +3/3 & -6/3 & +3/3 \end{bmatrix}$$

$$\text{Check } \left[\mathbf{A}\right]^{-1}\left[\mathbf{A}\right] = \left[\mathbf{I}\right] = \begin{bmatrix} 1 & 0 & 0 \\ 0 & 1 & 0 \\ 0 & 0 & 1 \end{bmatrix}$$

A.1.8   *System of equations*

From section 2.3.1 a solution procedure was given for three linear algebraic equations by an elimination procedure. Such a set of equations could be expressed in matrix form as

$$\left[\mathbf{A}\right]\{\alpha\} = \{\mathbf{B}\}$$

If the inverse of $\left[\mathbf{A}\right]$ had been found and multiplied into each side of the above equation

$$\left[\mathbf{A}\right]^{-1}\left[\mathbf{A}\right]\{\alpha\} = \left[\mathbf{A}\right]^{-1}\{\mathbf{B}\}$$

and remembering that $\left[\mathbf{A}\right]^{-1}\left[\mathbf{A}\right] = \left[\mathbf{I}\right]$

and that $\left[\mathbf{I}\right]\{\alpha\} = \{\alpha\}$

then $\{\alpha\} = \left[\mathbf{A}\right]^{-1}\{\mathbf{B}\}$ and the solution follows.

Finding the inverse of a matrix is very time-consuming and usually other methods of equation solutions are employed.

However, consider the equations

$$
\begin{array}{ccccccc}
x & + & 2y & + & 3z & = & 13 \\
4x & + & 5y & + & 6z & = & 32 \\
7x & + & 8y & + & 10z & = & 53
\end{array}
$$

or

$$
\begin{bmatrix} 1 & 2 & 3 \\ 4 & 5 & 6 \\ 7 & 8 & 10 \end{bmatrix} \begin{bmatrix} x \\ y \\ z \end{bmatrix} = \begin{bmatrix} 13 \\ 32 \\ 53 \end{bmatrix} \quad or \quad [\mathbf{A}]\{\alpha\} = \{\mathbf{B}\}
$$

$[\mathbf{A}]$ is as used in A.1.7 above and so from $\{\alpha\} = [\mathbf{A}]^{-1}\{\mathbf{B}\}$ and $[\mathbf{A}]^{-1}$ from above

$$
\{\alpha\} = \begin{bmatrix} x \\ y \\ z \end{bmatrix} = \begin{bmatrix} -2/3 & -4/3 & 1 \\ -2/3 & 11/3 & -2 \\ +1 & -2 & +1 \end{bmatrix} \begin{bmatrix} 13 \\ 32 \\ 53 \end{bmatrix} = \begin{bmatrix} 1 \\ 2 \\ 3 \end{bmatrix}
$$

A.1.9 *Matrix integration and differentiation*

Integration or differentiation of a matrix implies that each term is integrated or differentiated separately and the result located in the same position. Should the operation be applied to a matrix product, the matrix multiplication must be performed before the differentiation or integration.

A.2 Structural stiffness matrices

A.2.1 *More efficient storage*

Consider the simple pin-jointed structure shown in Fig. A.2.1. Each node has two degrees of freedom and will thus contribute two equations towards the final set of equations, the coefficients of which form the stiffness matrix. The uncondensed stiffness matrix will have the form shown below with x indicating possible non-zero term and blanks as zeros.

One immediately notices the symmetry about the diagonal terms and also how the terms cluster in a band about the diagonal. The horizontal breadth of this band is known as the bandwidth – in this case 15 terms. The half bandwidth which includes the diagonal term is 8. The number of zero terms is also significant:- 368 out of 576 – and so the matrix can be thought of as sparse and banded. In addition the stiffness matrices are usually symmetrical about the diagonal ($a_{ij} = a_{ji}$) and so the only data which need be stored is that within the area ABCD of the matrix – a total of 164 numbers. Storage would usually be achieved in a single array $[\mathbf{B}]$ 1×164 and not as a 24×24 array.

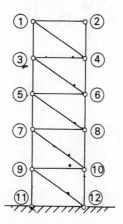

Figure A2.1

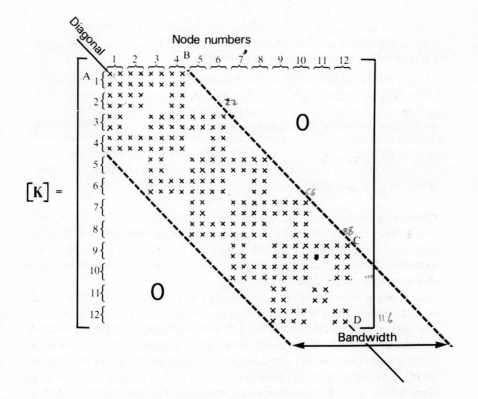

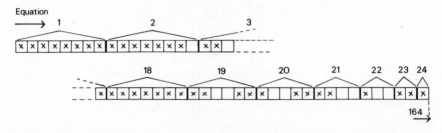

This 72% reduction in storage requirement however must be offset by the increase in programming complexity and this is the main reason this approach is not adopted in this book.

It is instructive to look at the form of the stiffness matrix of the structure in Fig. A.2.1 if it is renumbered as shown in Fig. A.2.2.

In this instance, whereas the density of non-zero terms is the same, the distribution is different and leads to a half bandwidth of 16. This illustrates the savings in storage that can be achieved by careful nodal numbering.

Footnote: In most large finite element packages will be found a storage scheme as described above in addition to an optimizer which will automatically renumber the structure to minimize bandwidth.

A.2.2 *Equation solution procedures*

The simple Gaussian elimination technique is used in the four programs presented in this book and no cognizance has been taken of the banding and symmetry of the stiffness matrix. A simple example of the procedure is given in section 2.3.1. Other procedures do exist and are used successfully in structural analysis, e.g. Gauss-Seidel iteration, triangular elimination (Banachiewicz-Crout), etc.

One other method, especially suitable for many finite element calculations, is the frontal solver, which is a progressive Gaussian elimination technique. Fuller details can be found in Hinton and Owen (see Appendix 1.1). If the numbering of a structure is ordered like that of Fig. A.2.1, the progressive build-up of the stiffness matrix element by element can mean that the top left of the stiffness matrix will be in the final form before all the elements have been processed. This means that concurrent with the establishing of the bottom right of the stiffness matrix, the Gaussian elimination of the top left can commence and proceed in a 'front' from top left to bottom right. The frontal solving routine thus finishes soon after the last element has been patched into the stiffness matrix.

Frontal solvers are ingenious and can save much storage and time but their programming is very complex and extensive. For this reason their inclusion in this book is not possible.

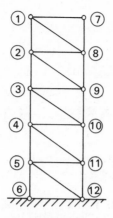

Figure 2.22

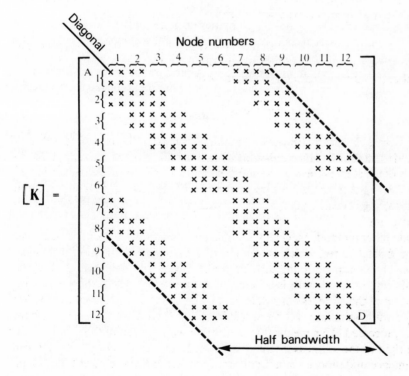

Appendix B BASIC – ours and yours*

B.1. Why BASIC?

A computer program is a set of instructions which direct the computer to perform a desired task. We may distinguish two broad classes of program – *systems programs,* written by professional programmers, which reside in the computer and undertake the multiplicity of tasks needed to allow the running of *user programs* such as the programs in this book.

In its innermost workings, the computer responds only to *binary signals* – in electrical terms these usually occur as two different voltage levels but from the programmer's point of view may be thought of as OFF/ON or HIGH/LOW or TRUE/FALSE and are most often represented by the binary digits 0 and 1. Computer programs written in this form are sometimes referred to as 'machine code'.

Programs (both systems and user) for the earliest computers were written in machine code but it soon became apparent that this 'language', while ideal for computers, was a sure route to ulcers, nervous breakdowns and worse for humans.

Among the substitutes invented for user programs were so-called 'high-level' languages which have a *vocabulary* of a small number of English words and a *syntax* – a set of rules for the construction of program statements. Hundreds of such languages have been devised – some for very specific purposes such as language translation or the control of astronomical telescopes, others for more general use. Programs written in these high-level languages have to be translated into machine code by a translator program resident in the computer. These translator programs are often large and very complex and their development may require many programmer-years of effort.

Two distinct types of translator program have been evolved – *interpreters*, which translate the source program as it is run on the computer, and *compilers*, which do a once-and-for-all translation of the source program into a so-called object program which is run. Each type of translator has its merits – an intepreter program editing and de-bugging (getting rid of errors) makes much simpler for the inexperienced user, but compiled programs run very much faster than interpreted programs.

*By D.A Pirie, B.Sc., M.S.

The high-level language used in this book, BASIC, originated at Dartmouth College, USA, in 1966. The name BASIC is an acronym for **B**eginners' **A**ll-purpose **S**ymbolic **I**nstruction **C**ode. The translator program for BASIC is nearly always an interpreter.

For any one of the multitude of uses to which computers may be put by experienced users, there is a better language than BASIC! Its value to the relative newcomer to computing lies not so much in the language itself as in the very simple operating environment in which it is usually found. By 'environment' we mean the set of procedures used for editing programs, for running programs and obtaining print-out of results, for storing programs and data on magnetic tape and/or disc and retrieving them therefrom; the environment is greatly influenced by whether the translator program is an interpreter or a compiler. On the modern desk-top microcomputer running an interpreted language, a reasonably numerate person can learn the essentials of these procedures in an hour or two.

By contrast, languages better suited to numerical computation, such as FORTRAN, are themselves more complex than BASIC. They are nearly always compiled and are usually surrounded by a much more elaborate operating environment. These factors place substantial extra barriers between the inexperienced user and the computer – and that is why BASIC has been used in this book.

B.1.1. *Our BASIC*

One of the weakness of BASIC which cannot be denied is the existence of so many different dialects (or versions) of the language. Section 2 of this Appendix describes briefly 'our' BASIC. This is 'Commodore BASIC version 2.0' (actually written by Microsoft) as implemented in the PET/CBM range of microcomputers. Actually, we have used only a subset of this BASIC (we have tried to avoid using features which might impede transfer of the programs to other microcomputers); our subset contains only a few features not found in standard 'minimal' BASIC.

We believe that only minor changes will be needed to allow the programs to be run on other microcomputers. We think we have separated out (into the handful of lines beginning at line 50000) those statements which will need changing. Some guidance on these modifications is given in Section 3 of this Appendix.

B.1.2. *Formatting of results*

Another major weakness of the versions of BASIC with which we are dealing is the almost complete lack of control offered over the format of printed results. We have gone to some trouble to overcome this deficiency by incorporating in the programs formatting subroutines (beginning at lines 10000 and 20000) which we would claim provide a degree of control of format comparable with (or better than) that offered by the PRINT USING command of more extensive BASICs or by the E and F formats of FORTRAN. Somme comments on these subroutines and guidance on their use are given in Section 4 of this Appendix.

B.2. Commodore BASIC 2.0

In this section we describe briefly only those features of Commodore BASIC which we have used in the programs in this book.

B.2.1 Data types

Computer programs manipulate data. In Commodore BASIC three different types of data are used – *integer, real* and *string* (or character). We have used only reals and strings. Reals are numbers within the range ±1.70141183E+38 (the E + 38 means 'times 10 to the power 38'); strings are sequences of keyboard characters (up to a maximum length of 255 characters).

B.2.2 Program elements — constants, variables, expressions

The above data types are used in *constants, variables* and *expressions.* Expressions are combinations of variables and constants (subject to the rules appropriate to the data type). It is worth keeping in mind that both *constants* and *variables* may be regarded as special cases of the more general category of *expression.*

Variable names may consist of one or two characters, the first being an uppercase letter $(A - Z)$ and the (optional) second being an uppercase letter or digit $(0 - 9)$. Names of more than two characters are tolerated but only the first two characters are significant. String variables are distinguished from reals by the addition of the $ sign.

X, XA and X1 are examples of permissible names for real variables. Examples of real constants are 1.234, 87, −0.0543. Real expressions are formed using the operations of exponentiation (raising to a power), multiplication, division, addition and subtraction. These operations must be indicated explicitly, using the symbols $\uparrow$, *, /, + and −; also, where necessary, parentheses (). An example of a real expression is $X \uparrow 2 + 2*X*Y* - Y/(LOG(Y) + 3)$; the LOG function illustrated here is one of a number of numerical functions 'built-in' to Commodore BASIC.

X$, XA$ and X1$ are examples of permissible names for string variables. Examples of string constants are "1.234", "FINITE ELEMENT METHOD", and "15TH SEPTEMBER 1983". The quotation marks are not part of the string. The only operation which may be performed on strings is that of *concatenation* (a fancy word for 'joining together'), the + sign being used to denote this operation. An example of a string expression is D$ +" " + LEFT$(M$,3) +" " + "1983"; the LEFT$ function illustrated here is one of a number of 'string-handling' functions in Commodore BASIC. If D$ and M$ were assigned the values "15TH"and "SEPTEMBER", respectively, the above expression would evaluate to "15TH SEP 1983". Strings and string-handling functions are used extensively in our output formatting subroutines.

Arrays of real, integer and string type may also be defined, using the same naming rule. Array variables have a totally separate existence from simple variables of the same name. Thus, for example, the real array A1() and the string array A1$() could be used in the same program as the simple real variable A1and the simple string variable A1$.

Commodore BASIC also allows *relational expressions* which take on the logical values TRUE or FALSE. These expressions make use of the relational operators ×, =, <, for example

$X > O$ (X is greater than O)
$MX < = MQ$ (MX is less than or equal to MQ)
$P < > Q$ (P is not equal to Q).

Relational expressions are made use of in control statements (see section B.2.3.4 below).

B.2.3 *Line-numbering, types of program statement*

BASIC programs are written as a sequence of numbered lines. The numbers increase from the beginning of the program to the end but any increments may be used between successive line numbers. Each line consists of one or more program statements, separated by a colon. The PET/CBM micros we used have a 40-column screen but allow a program line to have up to 79 characters (i.e. two screen widths, allowing for the <RETURN> terminator).

No matter how complex, a BASIC program actually contains only a very few different types of program statement. These are:
(i) assignment statements, which assign (give) a value to a variable;
(ii) input/output statements, which allow for the transfer of data;
(iii) control statements, which change the flow of execution within the program.

B.2.3.1. *Assignment statements.* This type of statement is the most common and is of the form

linenumber **LET** variable = expression

The variable and the expression must either both be numeric (real or integer) or must both be of string type. The statement means 'the value of the expression (on the right of the = sign) is assigned to the variable (on the left)'. The = sign in this context does not mean 'equals' – consideration of the perfectly valid statement

100 LET X = X + 1

should make that clear (it means 'assign to the variable X the value X+1' i.e. 'increment X by 1').

The LET keyword is actually optional in Commodore BASIC (as in the great majority of current BASICs) – we have omitted it throughout our programs.

B.2.3.2 *Input statements* – **INPUT, READ/DATA**. We have used two methods of inputting data to our programs –
(i) via the keyboard (i.e. the user) in response to INPUT statements and (ii) the READ statement which makes use of data embedded in the program in DATA statements.

The simplest form of INPUT statement is of the form

linenumber **INPUT** variable

When this statement is reached during program execution a
?

symbol appears on the screen to indicate to the user that an input from the keyboard is awaited. (Wherever we have used an INPUT statement we have also used a PRINT statement (see below) to put a rather more helpful prompt message on the screen). The <RETURN> key is used to indicate completion of the input.

If the variable is numeric, then so must be the input. If not, an error message

?REDO FROM START
?

will appear on the screen.

More than 1 value may be requested with a single INPUT statement of the more general form

linenumber **INPUT** variable, variable, ...,variable

for example

50 INPUT X, Y, Z

asks for 3 numeric values to be keyed-in. This may be done either as

constant,constant,constant< RETURN >

or as

constant < RETURN >
constant < RETURN >
constant < RETURN >

One annoying feature of INPUT is that the program will abort if the user accidentally presses the< RETURN > key before entering his data. We have adopted a primitive form of protection against this drop-out by arranging for a character to sit under the cursor when the ? prompt appears on the screen. This character is the asterisk * when numeric input is being asked for. When the user response is to be either Y (for YES) or N (for NO) the character is whichever is the more frequentley used response. This feature has been defined in a subroutine (beginning at line 53000) and so can easily be cancelled, if necessary.

The general form of the READ statement is very similar i.e.

linenumber **READ** variable, variable, ...,variable

Elsewhere in the program there must be one or more DATA statements containing the values (of appropriate data type) to be assigned to the variables specified in the READ statements. The DATA statement takes the form

linenumber **DATA** constant,constant, ...,constant

If the number of variables in the READ statement(s) exceeds the number of values contained in the DATA statement(s) then an

?OUT OF DATA

error message will appear on the screen.

B.2.3.3 *Ouput statements* –**PRINT** and **PRINT** #. Our programs make use of two output devices – the computer screen and a printer (at least 80 columns). The screen output consists of prompts and messages to the user and is produced by the PRINT statement. This is of the form

linenumber **PRINT** expression list

(remember that 'expression' includes 'variable' and 'constant' as special cases).

This statement is common to all BASICs. However, different BASICs differ in how they send data to a printer. Commodore BASIC, firstly, opens a channel to the printer with a command of the form

linenumber **OPEN** x,4

where x denotes an integer between 1 and 255. (For no good reason we can remember, we have used a value of 5 for x). Next, the data to be sent to the printer is included in PRINT # x statements, of the form

linenumber **PRINT #** x, expression list

Finally, the printer channel is closed with the statement

CLOSE x

Standard 'minimal' BASIC uses the comma and the semi-colon to separate items in the PRINT (or PRINT # x) expression list. Unfortunately, different BASICs and/or printers interpret the comma differently, in the resulting spacing of the items across the page. We have therefore adopted the device of assembling an entire line of printer output at a time (using only the string concatenation operator +). In this way we hope to achieve identical printer output with any printer and any BASIC capable of running the programs. To ease the pain of converting the programs to run on other micros we have written all the printer-related statements as subroutines (beginning at line 50000). Section 2.3.5 below describes briefly the main features of BASIC subroutines.

B.2.3.4 *Control statements* – **FOR . .NEXT, IF . .THEN, GOTO, ON . .GOTO.** In the absence of control statements, execution of a program would begin at the beginning (the lowest line number) and proceed to the end (the highest line number. In most programs, however, the need arises to repeat some part of the program several times (perhaps thousands) or to 'branch', i.e. go to different sections of the program depending on some criteria.

Let's look at the 'repeat a section of program' business first. The structure available in BASIC for this purpose is the **FOR . . .NEXT** loop. The FOR part is placed at the beginning of the block of programme to be repeated, and the NEXT part at the end. Thus, (using 'exp' to denote 'expression')

linenumber **FOR** real-variable = exp1 **TO** exp2 **STEP** exp3
statements to
be repeated
linenumber **NEXT** real-variable

(If exp3 = +1 then the STEP exp3 may be omitted.)

As a simple example consider the following segment of program

```
100 FOR K = 1 TO 3 STEP 0.5
110 PRINT K
120 NEXT K
```

This, when run, would produce the following output on the screen

```
1
1.5
2
2.5
3
```

after which execution would pass to the line following line 120 (if any).

Now let us turn our attention to the question of 'branching' – sending the flow of execution of the program in different directions. For this purpose we have the **IF . . . THEN** structure. In Commodore BASIC it takes the form

linenumber **IF** relexp **THEN** stat:stat: . . . . . :stat

(here we have used 'relexp' as an abbreviation for 'relational expression' and 'stat' for 'statement'). All of the statements after THEN are executed if the relational expression is TRUE and all are skipped if it is FALSE. (Minimal BASIC only allows one statement after THEN). In its simplest form there is only one 'stat', often just a transfer of execution to another line number, e.g.

```
170 IF X => 0.001 THEN GOTO 200
```

(the meaning of GOTO is fairly obvious!) Commodore BASIC allows this to be written

```
170 IF X => 0.001 THEN 200
```

GOTO also provides an unconditional transfer of control, in the form

linenumber1 **GOTO** linenumber2

Liberal use of the unconditional GOTO is an excellent way of making your programs incomprehensible to others. There may be some occasions when this is desirable but it is generally frowned on!

GOTO also features in the **ON . . . .GOTO** statement. This takes the form

linenumber **ON** exp **GOTO** linenumber1, linenumber2, . . .

There can be as many linenumbers after GOTO as will fit on the line. 'exp' is evaluated and truncated to an integer, if necessary. This integer must lie within the range 0 to 255 or the program will abort with

?ILLEGAL QUANTITY ERROR

The program branches to linenumber1 if exp = 1, to linenumber2 if exp = 2, etc. If exp=0 or has a value greater than the number of linenumbers (but < 256) no branch takes place. For example, in the following program segment

```
10 ON X GOTO 100,200
20 .....
100 ...
200 ...
```

the program will branch to line 100 if X = 1 and to line 200 if X= 2. If X = 0 or if 2 < X <=255 execution will move on to line 20.   If X < 0 or > 255 the program will abort.

B.2.3.5 *Subroutines* – **GOSUB, RETURN.** If this were a comprehensive account of BASIC then a much more prominent position would be given to the topic of subroutines.

As programs get bigger and more complex they become more difficult to write, debug and understand (even your own programs, 6 months later). The only effective antidote to this problem is to break the large program down into smaller units which can be developed and tested on their own. The subroutine has special properties which make it particularly effective in this regard.

Firstly, a subroutine can be 'called' from anywhere in the main program; it will then do its thing and return *to the statement immediately after the calling statement.* Subroutines can call other subroutines; in BASIC they can call themselves (this 'recursive' property is not shared by all computer languages).

Secondly, subroutines can be used in different programs. The discerning reader will have noted that all four programs in this book are identical from line 10000 onwards.

A weakness of all but few BASICs is that variables are 'global' – it is not, therefore, possible to use the same variable name independently in the main program and in a subroutine (unless you are very careful and+ or lucky!).

The 'call' statement to a subroutine beginning at linenumber1 takes the form

linenumber **GOSUB** linenumber1

The last statement of every subroutine must be

line number **RETURN**

B.2.3.6 *More about multiple statements per line.* This facility may be used to condense a program; if over done, program readability suffers. On the other hand, used in sympathy with the logical structure of a program, this feature is a *GOOD THING*. To give three examples:

(i) all the business of exchanging values of variables and parameters between the main program and a subroutine can often be taken care of on the same line (2 screen widths on the PET/CBM micros) as the GOSUB;

(ii) the statements contained within a short FOR . . . .NEXT loop may fit on one line and thereby be more easily comprehended;

(iii) all the statements which follow an IF< *relational expression*>THEN statement under the same line number are executed if the *relational expression* is TRUE and all are skipped if the *relational expression* is FALSE. This is a very useful extension to the IF . . .THEN statement of standard BASIC.

B.3 Your BASIC

If your micro is other than a 32K, 40-column PET/CBM you will naturally be wondering what would be involved in running our programs on your machine.

Clearly, we are not in a position to offer any cast-iron guarantees, but unless your micro is rare and exotic *it should be possible*. Furthermore, we think you will only need to change the handful of lines beginning at 50000 (where we have stowed all the business relating to use of the printer).

B.4 Use of the formatting subroutines

In common with most other smallish BASICs (occupying about 8K of memory), Commodore BASIC doesn't offer much help to users trying to produce neatly formatted output on a printer. We have tried to fill the gap with two subroutines, beginning at lines 10000 and 20000 respectively. Section B.2.3.5 contains a brief general description of subroutines.

Both subroutines do a formatting operation – lining up decimal points, rounding-off to a sensible number of figures and setting each number into a field of specified width. In addition, subroutine 10000 converts all numerical values to scientific notation (e.g. 1.234E + 05).

The two subroutines use the same names for the variables passed back and forth to the main programs, so it is a simple matter to switch from using one (at some point in a program) to the other – just change the appropriate GOSUB10000 to GOSUB20000 or vice-versa. The two subroutines may also be called with GOSUB10040 and GOSUB20040 respectively if values for the field width (FW) and number of digits (NS) other than the default values are to be used.

As noted earlier, these subroutines depend heavily on the string-handling functions LEFT$, RIGHT$ etc.

Index